D1165502

MECHANISMS OF EVERYDAY COGNITION

The West Virginia University Conferences on Life-Span Developmental Psychology

Datan/Greene/Reese: Life-Span Developmental Psychology: Intergenerational Relations
Cummings/Greene/Karraker: Life-Span Developmental Psychology: Perspectives on Stress and Coping
Puckett/Reese: Mechanisms of Everyday Cognition
Cohen/Reese: Life-Span Developmental Psychology: Methodological Contributions

MECHANISMS OF EVERYDAY COGNITION

Edited by
JAMES M. PUCKETT
Texas A & M University—Kingsville
HAYNE W. REESE
West Virginia University

LEA LAWRENCE ERLBAUM ASSOCIATES, PUBLISHERS
1993 Hillsdale, New Jersey Hove and London

Lawrence Erlbaum Associates, Inc., Publishers
365 Broadway
Hillsdale, New Jersey 07642

Library of Congress Cataloging-in-Publication Data

Mechanisms of everyday cognition / edited by James M. Puckett, Hayne W. Reese.
p. cm.
"Based on the proceedings of the twelfth biennial conference . . . at West Virginia University, held April 19–21, 1990"—Pref.
Includes bibliographical references and index.
ISBN 0-8058-0976-7
1. Cognition—Congresses. 2. Developmental psychology—Congresses. I. Puckett, James M. II. Reese, Hayne Waring, 1931–
BF311.M434 1993
153—dc20 93-10999
CIP

Contents

Contributors

Tina Blythe, Project Zero, Harvard Graduate School of Education, Cambridge, MA

Stephen J. Ceci, Department of Human Development, Cornell University, Ithaca, NY

Michael Chapman, deceased, late of Department of Psychology, University of British Columbia, Vancouver, BC

Paula Cuccaro, Department of Psychology, University of Houston, Houston, TX

William N. Dudley, Gerontology Center, University of Georgia, Athens, GA

Helene Hembrooke, Department of Psychology, SUNY Binghamton, Binghamton, NY

Frank H. Jurden, Department of Psychology, Boise State University, Boise, ID

Joseph S. Laipple, Department of Psychology, Saint Michael's College, Winooski, VT

Elizabeth F. Loftus, Department of Psychology, University of Washington, Seattle, WA

Lynn Okagaki, Department of Psychology, Yale University, New Haven, CT

Jeffrey T. Parsons, Department of Psychology, University of Houston, Houston, TX

Leslee K. Pollina, Department of Psychology, Southeast Missouri State University, Cape Girardeau, MO

Leonard W. Poon, Gerontology Center, University of Georgia, Athens, GA

James M. Puckett, Department of Psychology, Texas A & M University—Kingsville, TX

Hayne W. Reese, Department of Psychology, West Virginia University, Morgantown, WV

K. Warner Schaie, Department of Human Development and Family Studies, Pennsylvania State University, College Park, PA

Jonathan W. Schooler, Learning Research and Development Center, University of Pittsburgh, Pittsburgh, PA

Alexander W. Siegel, Department of Psychology, University of Houston, Houston, TX

Jan D. Sinnott, Department of Psychology, Towson State University, Towson, MD

Robert J. Sternberg, Department of Psychology, Yale University, New Haven, CT

Ruth H. Tunick, Department of Psychology, West Virginia University, Morgantown, WV

Richard K. Wagner, Department of Psychology, Florida State University, Tallahassee, FL

Julie Wall, Department of Psychology, University of Houston, Houston, TX

Joseph Walters, Project Zero, Harvard Graduate School of Education, Cambridge, MA

Armin D. Weinberg, Baylor College of Medicine, Baylor University, Waco, TX

Deborah J. Welke, Gerontology Center, University of Georgia, Athens, GA

Noel White, Project Zero, Harvard Graduate School of Education, Cambridge, MA

Sherry L. Willis, Department of Human Development and Family Studies, Pennsylvania State University, College Park, PA

Eugene Winograd, Department of Psychology, Emory University, Atlanta, GA

Preface

This volume is based on the proceedings of the 12th biennial conference on life-span developmental psychology at West Virginia University, held April 19–21, 1990, near Morgantown, West Virginia. Most of the contributions deal with the mechanisms of everyday cognition in one way or another, but a broad spectrum of additional concerns is addressed within the domain of everyday cognition: its metatheoretical underpinnings (see chapters by Michael Chapman; Jan Sinnott), theory and theoretical issues (see chapters by Robert Sternberg et al.; Sherry Willis and Warner Schaie), methods of investigation (see Leonard Poon et al.; Eugene Winograd), empirical considerations (see Alexander Siegel et al.; Jonathan Schooler and Elizabeth Loftus), and social issues and applications (see Stephen Ceci and Helene Hembrooke; Joseph Walters et al.).

Consistent with the life-span theme of this series, the contributors address everyday cognition in infancy (Ceci and Hembrooke), childhood (Walters et al.), adolescence (Siegel et al.), young and middle adulthood (Poon et al.; Sternberg et al.), and old age (Winograd; Willis and Schaie). The contributors also collectively address some of the traditional concerns of life-span psychology as they relate to everyday cognition across the chronological life span: the dialectical nature of everyday cognition (Chapman), individual differences (Schooler and Loftus), and contextual influences (Sinnott). Leading and concluding chapters provide overview, integration, and summary. In bringing together in this volume a wide array of age periods and points of view within the domain of everyday cognition, we hope that students and researchers in developmental psychology and cognitive science will find a useful cross-fertilization of ideas.

We would like to express appreciation for conference grant #1 R13 AG08999-01 from the National Institute of Aging (NIA). In particular, we thank

Robin Barr and Ron Abeles at NIA for their attendance and participation at the conference and for their support before, during, and after the conference. Also making the conference possible through their considerable and much appreciated contributions of time and effort were numerous doctoral students in psychology at West Virginia University: Mary Ballard, Mary Barnas, Edie Jo Hall, Frank Jurden, Joe Laipple, Leslee Pollina, Connie Toffle, and Ruth Tunick. We also acknowledge with gratitude the aid of the Department of Psychology and West Virginia University, which generously provided time and resources for the conference, and we thank Ann Davis for her excellent secretarial services. Finally, we thank the staff at Lawrence Erlbaum Associates for their patience and cooperation throughout the preparation of this book.

James M. Puckett
Hayne W. Reese

MECHANISMS OF EVERYDAY COGNITION

I INTRODUCTION AND OVERVIEW

1 An Integration of Life-Span Research in Everyday Cognition: Four Issues

James M. Puckett
Texas A & M University-Kingsville

Hayne W. Reese
Leslee K. Pollina
West Virginia University

When all of us involved in the planning of the conference on which this book is based came to the point of deciding what to name the conference and the book, we were faced with an everyday problem. How best to *market* the idea of a conference and book on life-span everyday cognition? How best to reach the intended audience? We debated whether to call the type of cognition in which we were interested "everyday" or "practical." Although "practical" had certain philosophical connotations that we wished to convey, "everyday" seemed to be the term that had become the most widely adopted. In the end, true to the spirit of our subject matter, pragmatic considerations outweighed idealistic ones in choosing the label *everyday.*

Both practical and theoretical issues are considered in the primary contributions to this volume and in the two integrative chapters. The final chapter, organized in terms of the ten primary contributions, is intended to summarize them while providing some integration among them. In this introductory chapter we focus on general themes in an attempt to provide an integration of the primary contributions with the larger literatures of everyday and laboratory cognition.

Among the recurring themes in the primary chapters are four that we consider here:

1. Is cognition best studied by controlled or naturalistic methods or some combination?
2. What differences exist between everyday/practical/real-world and laboratory/traditional/academic tasks and situations?
3. What are the mechanisms of practical cognition?

4. Do the mechanisms of everyday cognition differ from those of laboratory cognition?

METHOD AND THEORY IN EVERYDAY COGNITION

As might be suggested from the foregoing, the descriptors *everyday, practical,* and *real-world* are treated as being functionally equivalent for the purposes of this chapter. These terms are also treated as being more or less opposite the functionally equivalent descriptors *academic, laboratory,* and *traditional.* The members within each triad appear to be used interchangeably by many researchers.

Whatever labels we may choose to apply to them, the domains of practical and laboratory cognition have been dichotomized largely on two related bases: methodological and theoretical. The methodological basis for distinguishing between everyday and laboratory cognition, of course, concerns the relative merits of naturalistic vs. controlled observation (Intons-Peterson, 1992; Loftus, 1991; Poon, Welke, & Dudley, this volume; Winograd, this volume). In referring to the theoretical basis, we mean dimensions of tasks or dimensions of contexts that differentially characterize the real world and the laboratory (e.g., Chapman, this volume; Puckett, Reese, Cohen, & Pollina, 1991; Sinnott, 1989a; Willis & Schaie, this volume). In the evolution of research on everyday cognition, as we explain shortly, these bases appear to have played different roles in life-span psychology and in the larger domain of cognitive psychology.

However, we must be careful not to ascribe too much validity or significance to the method-theory distinction. Many researchers calling for naturalistic methodology do so in hopes of uncovering new theoretical principles (e.g., Neisser, 1991; Winograd, this volume). Likewise, researchers focusing on the theoretical properties of practical cognition (e.g., Siegel, Cuccaro, Parsons, Wall, & Weinberg, this volume; Sternberg, Wagner, & Okagaki, this volume; Walters, Blythe, & White, this volume) tend to utilize somewhat more naturalistic research settings and techniques than do strictly traditional laboratory researchers. That is, the methodological and theoretical bases may not be truly independent in the context of everyday cognitive research. The method-theory distinction, however, does provide one of many possible conceptual frameworks that might help organize the research domain of everyday cognition.

What Methods are Best for Studying Everyday Cognition?

For cognitive scientists in general, it can be argued that the methodological schism between practical and laboratory cognition is primary and that theoretical distinctions are secondary. The origin of current interest in everyday cognition appears to be a paper in which Neisser (1978) attacked the laboratory approach that emphasizes internal validity over external validity, charging that nothing interesting or important had resulted from roughly 100 years of effort in the laboratory. That paper and reactions (e.g., Banaji & Crowder, 1989) and counter-reactions (e.g., Ceci & Bronfenbrenner, 1991; Neisser, 1991) to it continue to drive interest and controversy in this area (e.g., Banaji & Crowder, 1991; Intons-Peterson, 1992).

In response to Neisser (1978), Banaji and Crowder (1989) defended the laboratory approach to cognition, asserting that whenever the goals of external validity and internal validity conflict, external validity should be sacrificed. This conflict concerning the role of different types of validity can be understood with reference to world views. Hultsch and Hickey (1978) argued that the relative weighting of external validity, internal validity, and the other types of validity identified by Cook and Campbell (1975) is different in different world views. Specifically, Hultsch and Hickey argued that the logical sequence of concerns in mechanistic perspectives begins with statistical conclusion validity and proceeds to internal validity, then construct validity, and finally external validity. In dialectical perspectives (organismic and contextualistic, as defined by Pepper, 1942), the logical sequence is reversed. Banaji and Crowder's (1989, 1991) position is consistent with the mechanistic perspective, whereas the primacy of external validity over internal validity is consistent with the dialectical perspective that has been explicitly adopted by many everyday researchers (e.g., Chapman, this volume; Sinnott, this volume). Thus, the debate about the relative importance of external and internal validity is likely to be futile unless the world view orientations of the debaters are explicitly identified (see also Poon et al., this volume, for a discussion of the role that misunderstanding of different types of validity has played in this debate).

Although Banaji and Crowder (1989) did acknowledge that external validity is desirable where feasible, they also declared that the everyday movement in memory research was bankrupt. Ceci and Bronfenbrenner (1991), Gruneberg, Morris, and Sykes (1991), Loftus (1991), Neisser (1991), Winograd (this volume) and others have taken issue with this characterization, defending the everyday approach with examples of useful research and expressing concerns with Banaji and Crowder's rationale. Continuing the exchange, Banaji and Crowder (1991) and Roediger (1991) reiterated and expanded on points made by Banaji and Crowder (1989). Roediger (1991) aptly applied a metaphor from current popular music in pointing out that "We [the laboratory methodologists] didn't start the fire." The relevant point for our present purposes, of course, is not who started the fire, but that it is still burning and that it was based first upon method, not theory.

Despite the divisive rhetoric, a consensus appears to be emerging that both naturalistic and controlled methods should be used in the study of cognition (Banaji & Crowder 1991; Ceci & Bronfenbrenner, 1991; Intons-Peterson, 1992;

Landauer, 1989; Neisser, 1988; Poon et al., this volume; Roediger, 1991; Rubin, 1989; Winograd, this volume). The differences of opinion that remain appear to lie in the relative emphasis placed on each methodological approach. For example, on one end of this continuum, Winograd (this volume) provides numerous examples of research that has been conducted without traditional experimental controls, although he simultaneously advocates the use of laboratory methods and the use of laboratory-derived principles in explaining findings from the real world. On the other end of the continuum, Willis and Schaie (this volume) suggest the use of primarily traditional psychometric and controlled laboratory methods, although they also mention naturalistic observation.

Yet, despite a verbal commitment to the use of both controlled and naturalistic methods by an apparent majority of laboratory and everyday researchers, Poon et al. (this volume) demonstrate with a meta-analysis of everyday and laboratory research that most researchers in both camps utilize primarily controlled methods. Indeed, with the exception of Winograd (this volume), all the contributors present either primary or secondary analyses of data derived largely or wholly with the use of controlled methods. Ceci and Hembrooke (this volume), Chapman (this volume), Schooler and Loftus (this volume), Sinnott (this volume), and Willis and Schaie (this volume) present theoretical integrations of data collected under mainly controlled conditions. Poon et al. (this volume), Siegel et al. (this volume), Sternberg et al. (this volume), and Walters et al. (this volume) present more detailed descriptions of empirical studies in which the data were also collected under mainly controlled conditions.

In summary, there is some difference of opinion (apparently arising from differences in world views) concerning the relative emphasis to be placed on controlled and naturalistic approaches, but almost all researchers from both camps agree that both approaches should be used. Meanwhile, most psychologists who study cognition seem mainly to be using controlled methods.

What Theoretical Dimensions Distinguish Everyday and Laboratory Contexts?

A second basis of distinction between practical and laboratory cognition is theoretical. Theoretical distinctions between real-world and laboratory tasks seem not to have played a primary role in shaping the debate in cognitive psychology generally. Theoretical differences, however, appear to have played a prominent role in life-span research. Theoretical dimensions have been said to distinguish the tasks and contexts in the real world from the tasks and contexts in the laboratory (e.g., Arlin, 1989; Sinnott, 1989b), although not all researchers agree that these distinctions are necessarily warranted (e.g., Puckett et al., 1991). In any case, theoretical concerns seem to have played a more important role than have methodological concerns for many researchers within life-span developmental psychology.

To illustrate this point, old adults have consistently been found to perform more poorly on academic and laboratory cognitive tasks than do young adults (e.g., Kausler, 1991; Reese & Rodeheaver, 1985). As outlined by Sinnott (1989a), many gerontological researchers became increasingly convinced that this poorer performance was not attributable to a deficiency in overall cognitive abilities but rather resulted from difficulties in performing laboratory tasks, difficulties that would not be observed in real-world situations. Thus, the search began for the defining dimensions of laboratory and academic tasks and situations that would distinguish them from everyday tasks and situations.

As examples of the theoretical distinctions between the practical and laboratory domains, Sinnott (1989b) held that everyday tasks, as opposed to laboratory tasks, are more frequent ("Is this ever likely to happen to anyone,") and significant ("and if it did, would they care?"). Ceci and Hembrooke (this volume) emphasize the greater contextual richness and variability of real-world tasks. Willis and Schaie (this volume) also assert that a consensus exists among most researchers that certain aspects of problems such as greater task complexity and multidimensionality characterize real-world tasks, but they assert that agreement is not widespread on certain other dimensions. Some of these other dimensions, including ill- vs. well-defined, are nevertheless recognized by Sternberg et al. (this volume) and by Chapman (this volume). Walters et al. (this volume; see also Walters & Gardner, 1986) enumerate a number of additional characteristics of problems, for example, interpersonal and intrapersonal, that tend to describe everyday tasks more than they do academic ones.

Therefore it appears that for many life-span researchers in practical cognition, the theoretical dimensions that define the everyday and laboratory domains continue to be important with a significant diversity of opinion prevailing as to what these dimensions are. Perhaps the diversity exists because there has been little investigation of these dimensions (see Puckett et al., 1991, for additional review). This situation is in contrast to that for methodology, as described earlier, for which consensus is fairly high regarding the methods that should be used (i.e., both controlled and naturalistic) and regarding the methods that are actually used (i.e., mainly controlled; Poon et al., this volume).

MECHANISMS IN EVERYDAY AND LABORATORY COGNITION

Related to the issue of task dimensions just considered is the issue of the mechanisms underlying performance on everyday and laboratory tasks. As noted earlier, for example, some researchers propose that certain descriptions characterize real-world more than laboratory tasks, such as greater complexity of tasks and contexts (Ceci & Hembrooke, this volume; Willis & Schaie, this volume). Regardless of whether such descriptions are true of everyday tasks, a separate issue concerns which cognitive mechanisms underlie performance in real-world tasks. And yet another issue is whether the same cognitive mechanisms that are brought to bear in, for example, purportedly complex and rich real-world contexts are also brought to bear in purportedly more simple and sterile laboratory contexts.

Salthouse (1991) outlined six overlapping levels of analysis of cognitive phenomena that will be helpful in delimiting the scope of our discussion about mechanisms. These levels are world views, frameworks, theories, models, descriptive generalizations, and empirical observations. A *world view* (e.g., mechanistic, organismic, or contextualistic) encompasses often implicit assumptions about the way organisms function. *Frameworks* are loose collections of stated assumptions and concepts (e.g., associationism, information processing, and life-span frameworks). *Theories* are explicitly stated relations among concepts (e.g., Chapman, this volume). *Models* specify the relations between theory and empirical observations in terms of mechanisms, although one can theorize without the constraints of a formal theory regarding the identity and operation of mechanisms (e.g., Marx, 1976). *Descriptive generalizations* are summaries of empirical observations in terms of, say, a mathematical equation. Finally, *empirical observations* are closely tied to, but abstractions of, actual behavior.

As examples of these levels of analysis employed in this volume, Sinnott engages in discourse primarily at the levels of methatheory and framework, discussing what practical cognition can gain from examining assumptions shared with other sciences. Willis and Schaie present a life-span framework and associated mechanisms. Chapman briefly discusses metatheory, but also elaborates a theory of everyday reasoning. Sternberg et al. and Walters et al. develop and test restricted applications of their respective theories using what might be called models. Several other contributors (Ceci and Hembrooke; Schooler and Loftus; Siegel et al.) propose mechanisms at the level of modeling, although usually without reference to a specific theory. Winograd and Siegel et al. present descriptive generalizations based on the behavior of old adults and adolescents, respectively. Most of the contributors engage in multiple levels of analysis, but in this section we are primarily concerned with analyses at the level of theories and models, because these are the levels at which mechanisms are proposed and tested.

What Are the Mechanisms of Everyday Cognition?

Cognitive mechanisms that have been proposed to mediate practical cognitive performance can be considered from a number of perspectives. In this section, we consider mechanisms in terms of whether they are serial (e.g., Anderson, 1983) or parallel (e.g., McClelland & Rumelhart, 1986; Rumelhart & McClelland, 1986). One particular set of mechanisms of practical cognition, those associated with the creation and use of domains of expertise, is explored in some depth. Finally, we explore whether certain everyday mechanisms are, or can be, used to account for change across the life span.

Serial and Parallel Processing. The serial-parallel distinction is important in life-span research because of the roles of both effortful, serial processing and automatized, parallel processing in theorizing at every stage of the life span (e.g., Case, 1985; Pascual-Leone, 1983). Generally speaking, automatization transforms the processing of information from a serial (effortful) mode to a parallel (automatized) mode; the end result is relatively effortless expertise in an area. The mechanisms involved in automatization have been studied and modeled in elaborate detail (e.g., Anderson, 1983).

Ceci and Hembrooke (this volume) speak of the importance, if not the predominance, of infants' processing of such amodal types of information as familiarity (or frequency). Such presumably genetically based automaticity has also played a role in cognitive aging research (e.g., Kausler & Puckett, 1980).

Automaticity of the environmentally based (overlearning) type (e.g., Shiffrin & Schneider, 1977) is used by Winograd (this volume) to explain an everyday memory phenomenon in adults. In the experiential subtheory of his triarchic theory, Sternberg (1985) argued for a relationship between experience and insight. Wisdom in old adults (e.g., Baltes & Smith, 1990) might be considered akin to experience and insight and is sometimes considered a form of automatized, parallel processing (Pascual-Leone, 1983). Sinnott (this volume) includes wisdom as one of the ultimate goals of postformal cognitive growth. In short, various forms of parallel processing are used in life-span theorizing.

Much of the theorizing in this volume is about mechanisms of the serial type (e.g., Schooler and Loftus). We consider some of these mechanisms shortly. Nevertheless, most of the contributors would probably acknowledge that even the serial mechanisms that are presented in this volume would become parallel given enough practice (Anderson, 1983; Shiffrin & Schneider, 1977), creating what Rybash, Hoyer, and Roodin (1986) might call domains of everyday expertise.

Domains of Expertise. Rybash et al. (1986) proposed that cognition is "encapsulated" within domains of expertise, viewing different areas of everyday functioning as independent content areas, each either being or failing to be a domain of expertise. These areas of expertise may consist of declarative knowledge (e.g., vocabulary definitions), procedural knowledge (e.g., word-processing), or some combination (Anderson, 1983). Viewing practical cognition in terms of domains of expertise appears to be common (e.g., Ceci & Liker, 1986).

If one grants that a domain of expertise consists of knowledge that has become automatized and must therefore be accessed through parallel search, then the mechanisms of automatization and parallel search are in the domain of everyday cognition. As such, they appear to be implicitly or explicitly accepted by most of the present contributors. Willis and Schaie implicitly refer to these mechanisms in referring to knowledge (of the declarative type) and skills (procedural knowledge) as components of practical intelligence. Chapman discusses the procedural knowledge domains of everyday and formal reasoning. Sternberg et al. study domains of expertise, which they call tacit knowledge, for managers, academicians, and salespeople, and they investigate the trainability of tacit knowledge in schools. Walters et al. also investigate the trainability of different areas of practical intelligence in the schools. Thus, domain-specific expertise is a generally accepted construct in theorizing about everyday cognition.

Some of the contributors differ, however, in whether or not they acknowledge a domain-general component of practical cognition. Sternberg et al. (this volume) find evidence for a general or "g" factor in practical intelligence in that different areas of tacit knowledge correlate reliably with each other. Therefore, they propose both domain-specific and domain-general components. In contrast, Walters et al. (this volume; Gardner, 1983; Walters & Gardner, 1986) maintain that each area of intelligence is specific and independent.

Life-Span Cognitive Change Mechanisms. Whether or not any given mechanism operates on a day-to-day basis in producing everyday cognitive performance, a separate issue is whether changes over the life span involving that mechanism are responsible for changes in observable cognitive performance (age-related differences) across the life span. One can explain *why* the age-related differences occur and *how* the age-related differences occur. Salthouse (1991) said the former is a function of a developmental theory, and the latter he called a change mechanism.

The present contributors offer a variety of mechanisms that might be involved in life-span changes in cognitive performance. Perhaps the most prevalent of these is the development of domains of expertise through the automatization of knowledge. Increasing expertise in areas of practical cognition across the life span helps account for the cognitive growth that Sinnott investigates, the crystallized intelligence studied by Willis and Schaie, and the process of sociogenesis to which Chapman refers. For example, in Chapman's theory, as external argumentation procedures become internalized through sociogenesis, their transformation toward covertness and expertness conforms to the well-known and studied sequence that characterizes the transformations of automatization (e.g., Shiffrin & Schneider, 1977). A question that merits investigation is whether the process of sociogenesis for reasoning also leads to other consequences known to result from automatization such as difficulty in suppressing responses and difficulty in consciously inspecting mental activities (Shiffrin & Schneider, 1977).

Other principles proposed by the contributors to this volume also have promise in accounting for life-span changes in everyday cognition. Ceci and Hembrooke suggest that infants may actually encode more information than do older children, but less selectively. Perhaps the overly unselective encoding occurs because infants engage in more data-driven processing (Roediger & Blaxton, 1987) than do adults. This trend away from data-driven toward conceptually driven processing may continue across the life span, with old adults relying on conceptually driven processing more than do young adults. Many theorists and findings pointing toward increased reliance on previously learned crystallized experience in old age would support such a life-span trend (e.g., Pascual-Leone, 1983).

This mechanism would help explain infants' tendency (Ceci & Hembrooke, this volume) to be more sensitive to changes in context than are older children, by virtue of the rationale that infants would encode meaningless, less predictive features of the environment that when altered on a later occasion would disrupt memory retrieval. Could a shift from relatively data-driven toward more conceptually driven processing also explain age-related differences in selectivity of encoding in adulthood? It is fairly well established that old adults' performance is more disrupted by distractions in the environment than that of young adults (e.g., Park, 1992). This greater distractability may represent a tendency toward less selective encoding in old adults that might be explained by a trend toward more conceptually driven processing. For example, old adults could be so conceptually driven that through disuse (or deterioration) they have lost the ability to selectively attend and encode.

Selectivity of retrieval (and possibly also encoding) also appears to be involved in changes in eyewitness memory performance across the life span (Schooler & Loftus, this volume). Young adults are found to exhibit better eyewitness performance than do children and old adults. Schooler and Loftus invoke selectivity of retrieval in explaining the observation that older adults are more likely to recall misinformation than are young adults. Older adults appear to accept and encode misinformation unselectively, but their retrieval on a later memory test may be selective in that the content of the misinformation but not its source is remembered. Young children are also more likely to accept misinformation than are adults, perhaps reflecting less selectivity of encoding.

The contributors offer other mechanisms of practical cognition that need not be summarized here. Some of these mechanisms that are posited only for a given age period, however, might upon examination be found to be involved in changes in cognition across the life span. Whether or not they prove useful as change mechanisms or components of change mechanisms, the variety of mechanisms offered by the contributors indicates that the area of everyday cognitive research is in a healthy, nonbankrupt condition.

Do Everyday Mechanisms Differ From Laboratory Mechanisms of Cognition?

In the previous section, we considered whether the knowledge and mechanisms of everyday cognition were specific or general within the domain of everyday cognition as a whole. In this section, we consider a similar question on a larger scale, namely, to what extent are mechanisms of everyday cognition specific to the everyday domain or shared with those of laboratory cognition? We then consider what each area of research can gain from an integration with the other area. Finally, we briefly review theoretical and metatheoretical developments intended to integrate the domains of practical and laboratory cognition.

Shared or Separate Mechanisms? Willis and Schaie (this volume) conceptualize practical cognition as being hierarchically related to other types of cognition. Genotypic intelligence consisting of basic abilities can be placed at the top of the hierarchy, and subsets of abilities and knowledge can be viewed as organized to meet the needs of each task or domain. Everyday and laboratory cognition might therefore be viewed as separate manifestations of intelligence that have been organized to meet the demands of the respective contexts. This hierarchical view of intelligence would seem to require that practical and laboratory cognition utilize the same mechanisms, even if they are organized differently for each domain.

A hierarchical view also seems to have been held by Sternberg (1985), whose contextual subtheory of the triarchic theory depicts adaptation as selecting, shaping, or adapting to different environments. In adapting to these different environments, elements of the componential subtheory appear to be organized to meet the needs of each particular context. For example, tacit knowledge (Sternberg et al., this volume) consists of knowledge and skills organized for different contexts.

Chapman (this volume) theorizes that the same process of sociogenesis is responsible for the development of everyday and formal reasoning, but he points out that formal reasoning may require additional attentional capacity to be effectively executed. Chapman's theory might be considered another example of a hierarchical theory.

Walters et al. (this volume; Gardner, 1983) hold a different view of intelligence that might be described as horizontally organized rather than vertically or hierarchically organized. Laboratory and academic tasks are thought to require primarily linguistic and mathematical intelligence, whereas practical cognition is viewed as being composed not only of these two but also intrapersonal, interpersonal, and all the other types of intelligence. This theory seems to imply that the mechanisms of laboratory and practical cognition are shared only to the extent that domains of intelligence are shared. In some cases, a given practical and laboratory task will have identical mechanisms, and in some cases they may have completely independent mechanisms, depending on the intelligences or combination of intelligences tapped by the particular tasks.

Sinnott (this volume) also appears to assume a horizontal organization of intelligence. She states that the mechanisms of practical cognition are similar to, but different from, those of information processing. Among the everyday mechanisms are what she calls strange loops, which appear to be akin to what Hayes-Roth and Hayes-Roth (1979) call opportunistic planning.

A third view seems to be shared by a majority of the present contributors, namely, that the same mechanisms are used in everyday and laboratory cognitive tasks. This position was perhaps best captured by Leonard Poon at the conference on which this book is based, in the heuristic "Research is research is research, and cognition is cognition is cognition." Similarly, Winograd (this volume) proposes laboratory-based principles (that hint of laboratory-derived mechanisms) to explain findings from research conducted with naturalistic methods in the real world. Ceci and Hembrooke (this volume) utilize findings from animal laboratory studies and from field and laboratory studies of humans, suggesting mechanisms to explain infant memory that are of the same type used to explain laboratory cognition. Siegel et al. (this volume) discuss everyday cognition in terms of neo-Piagetian mechanisms that have been applied to both laboratory and real-world situations by Piaget and others (e.g., Elkind, 1985). Finally, Schooler and Loftus (this volume) theorize in terms of laboratory-derived mechanisms in order to understand eyewitness memory. For all of these contributors, the mechanisms of everyday and laboratory cognition appear to be indistinguishable. Considering all the proponents of the vertical (hierarchical), horizontal, and cognition-is-cognition positions, then, the predominant view is that the same mechanisms are utilized in practical and laboratory tasks.

What Can the Areas Gain From Each Other? As some authors in this volume (e.g., Poon et al.) and elsewhere (e.g., Ceci & Bronfenbrenner, 1991) argue, laboratory research can enrich and inform everyday research and vice versa. Winograd (this volume), for example, argues explicitly that principles derived in the laboratory can be used to explain empirical findings obtained in the real world. Utilization of laboratory-derived mechanisms to explain everyday research findings is common also beyond the research found in this volume (e.g., Pollina, Greene, Tunick, & Puckett, 1992, in press). In a fashion similar to that of Siegel et al. (this volume), Pollina et al. (1992, in press) collected self-ratings of everyday cognitive performance, analyzed the ratings with factor analysis, used laboratory-derived cognitive constructs to describe the factors, and finally, similarly to Schooler and Loftus (this volume), Pollina et al. (1992) explored whether individual differences in self-ratings predicted actual cognitive performance. In the current volume, other examples of this directional influence from the laboratory to the practical domain (e.g., Schooler and Loftus on eyewitness memory) are in evidence.

Equally importantly, everyday research can enrich and inform laboratory research. Several authors in this volume (e.g., Winograd) and elsewhere (e.g., Bahrick, 1991) make the case that naturalistic observation in the real world is the best way to describe phenomena, and everyday research is the best way to test the validity of the principles derived in the laboratory. Puckett et al. (1991) outlined a framework whereby the results of everyday research can help to resolve a longstanding debate concerning laboratory research on young and old adults. The debate centers on the best explanation for age-related declines in performance on laboratory tasks. Do the apparent declines in the performance of old adults represent biologically based declines in abilities, disuse of academic strategies, or some other explanation? Analyses of performance (e.g., solution adequacy) and strategies along various dimensions of tasks said to differentiate real-world and laboratory tasks (e.g., ill- vs. well-defined) may reveal patterns of results that help to determine which explanation is correct for each task dimension. Early results using this framework suggest that a disuse explanation may be supported for the familiarity dimension, although different explanations may emerge for different task dimensions tested.

Integrating Theory and Research in Everyday and Laboratory Cognition. Although research in practical cognition that is not guided by a particular theory is valuable in its own right, Salthouse (1991) maintained that theory-driven research can potentially lead to even more rapid strides in understanding. A large scale theoretical integration involving the theories of Sternberg (1985) and Gardner (1983) has begun, each of which is broad enough to encompass both everyday and academic cognition. In conjunction with the Practical Intelligence for Schools (PIFS) program (Sternberg et al., this volume; Walters et al., this volume), Sternberg's process-oriented triarchic theory is "infused" into Gardner's content-oriented theory of multiple intelligences in order to design curriculum units to assess whether practical intelligence can be taught to school children. If these theories are successfully integrated, the integrated theory will have great promise in furthering exploration of the mechanisms of everyday and laboratory cognition.

Puckett et al. (1991) have argued that a comprehensive theory of this type should also be integrated with a computer simulation theory such as that of Anderson (1983) in order to facilitate the analysis of mechanisms. Craik and Jennings (1992) suggested that an integration of cognitive aging research with the areas of neuropsychology and cognitive neuroscience would also accelerate understanding of adult cognitive development. We think that this integration would equally benefit life-span cognitive research.

Sinnott (this volume) provides a glimpse of an overarching framework that should encourage and facilitate such integrations. Her framework ties together concepts from General Systems Theory (e.g., Laszlo, 1972) and concepts from several other theories and sciences including chaos theory, the new biology, the new physics, and postformal cognitive psychology. All of these prospects make this an exciting time for research in cognition.

REFERENCES

Anderson. J. R. (1983). *The architecture of cognition.* Cambridge, MA: Harvard University Press.

Arlin, P. K. (1989). The problem of the problem. In J. D. Sinnott (Ed.), *Everyday problem solving: Theory and applications* (pp. 229–237). New York: Praeger.

Bahrick, H. P. (1991). A speedy recovery from bankruptcy for ecological memory research. *American Psychologist, 46,* 76–77.
Baltes, P. B., & Smith, J. (1990). Toward a psychology of wisdom and its ontogenesis. In R. J. Sternberg (Ed.), *Wisdom: Its nature, origins, and development* (pp. 87–120). New York: Cambridge University Press.
Banaji, M. R., & Crowder, R. G. (1989). The bankruptcy of everyday memory. *American Psychologist, 44,* 1185–1193.
Banaji, M. R., & Crowder, R. G. (1991). Some everyday thoughts on ecologically valid methods. *American Psychologist, 46,* 78–79.
Case, R. (1985). *Intellectual development: Birth to adulthood.* Orlando, FL: Academic Press.
Ceci, S. J., & Bronfenbrenner, U. (1991). On the demise of everyday memory. "The rumors of my death are much exaggerated" (Mark Twain). *American Psychologist, 46,* 27–31.
Ceci, S. J., & Liker, J. (1986). Academic and nonacademic intelligence: An experimental separation. In R. J. Sternberg & R. K. Wagner (Eds.), *Practical intelligence: Nature and origins of competence in the everyday world* (pp. 119–142). New York: Cambridge University Press.
Cook, T. C., & Campbell, D. T. (1975). The design and conduct of quasi-experiments and true experiments in field settings. In M. D. Dunnette (Ed.), *Handbook of industrial and organizational research* (pp. 223–326). Chicago: Rand McNally.
Craik, F. I. M., & Jennings, J. M. (1992). Human memory. In F. I. M. Craik & T. A. Salthouse (Eds.), *The handbook of aging and cognition* (pp. 51–110). Hillsdale, NJ: Lawrence Erlbaum Associates.
Elkind, D. (1985). Egocentrism redux. *Developmental Review, 5,* 218–226.
Gardner, H. (1983). *Frames of mind: The theory of multiple intelligences.* New York: Basic Books.
Gruneberg, M. M., Morris, P. E., & Sykes, R. N. (Eds.). (1991). The obituary on everyday memory and its practical applications is premature. *American Psychologist, 46,* 74–76.
Hayes-Roth, B., & Hayes-Roth, F. (1979). A cognitive model of planning. *Cognitive Science, 3,* 275–310.
Hultsch, D. F., & Hickey, T. (1978). External validity in the study of human development: Theoretical and methodological issues. *Human Development, 21,* 76–91.
Intons-Peterson, M. J. (1992). Brief introduction to the special issue on memory and cognition applied. *Memory and Cognition, 20,* 323–324.
Kausler, D. H. (1991). *Experimental psychology, cognition, and human aging* (2nd ed.). New York: Springer-Verlag.
Kausler, D. H., & Puckett, J. M. (1980). Frequency judgments and correlated cognitive abilities in young and elderly adults. *Journal of Gerontology, 35,* 376–382.
Landauer, T. K. (1989). Some bad and some good reasons for studying memory and cognition in the wild. In L. W. Poon, D. C. Rubin, & B. A. Wilson (Eds.), *Everyday cognition in adulthood and late life* (pp. 116–125). Cambridge, UK: Cambridge University Press.
Laszlo, E. (Ed.). (1972). *The relevance of general systems theory: Papers presented to Ludwig von Bertalanffy on his seventieth birthday.* New York: George Braziller.
Loftus, E. F. (1991). The glitter of everyday memory . . . and the gold. *American Psychologist, 46,* 16–18.
Marx, H. H. (1976). Theorizing. In M. H. Marx & F. E. Goodson (Eds.), *Theories in contemporary psychology* (2nd ed., pp. 261–286). New York: Macmillan.
McClelland, J. L., & Rumelhart, D. E. (1986). *Parallel distributed processing: Vol. 2. Psychological and biological models.* Cambridge, MA: MIT Press.
Neisser, U. (1978). Memory: What are the important questions? In M. M. Gruneberg, P. E. Morris, & R. N. Sykes (Eds.), *Practical aspects of memory* (pp. 3–24). San Diego, CA: Academic Press.
Neisser, U. (1988). New vistas in the study of memory. In U. Neisser & E. Winograd (Eds.), *Remembering reconsidered: Ecological and traditional approaches to the study of memory* (pp. 1–10). Cambridge, UK: Cambridge University Press.
Neisser, U. (1991). A case of misplaced nostalgia. *American Psychologist, 46,* 34–36.

Park, D. C. (1992). Applied cognitive aging research. In F. I. M. Craik & T. A. Salthouse (Eds.), *The handbook of aging and cognition* (pp. 449–493). Hillsdale, NJ: Lawrence Erlbaum Associates.

Pascual-Leone, J. (1983). Growing into human maturity: Toward a meta-subjective theory of adulthood stages. In P. B. Baltes & O. G. Brim (Eds.), *Life-span development and behavior* (Vol. 5, pp. 117–156). New York: Academic Press.

Pepper, S. C. (1942). *World hypotheses: A study in evidence.* Berkeley: University of California Press.

Pollina, L. K., Greene, A. L., Tunick, R. H., & Puckett, J. M. (1992). Dimensions of everyday memory in young adulthood. *British Journal of Psychology, 83,* 305–321.

Pollina, L. K., Greene, A. L., Tunick, R. H., & Puckett, J. M. (in press). Dimensions of everyday memory in late adulthood. *Current Psychology.*

Puckett, J. M., Reese, H. W., Cohen, S. H., & Pollina, L. K. (1991). Age differences versus age deficits in laboratory tasks: The role of research in everyday cognition. In J. D. Sinnott & J. C. Cavanaugh (Eds.), *Bridging paradigms: Positive development in adulthood and cognitive aging* (pp. 113–130). New York: Praeger.

Reese, H. W., & Rodeheaver, D. (1985). Problem solving and complex decision making. In J. E. Birren & K. W. Schaie (Eds.), *Handbook of the psychology of aging* (2nd ed., pp. 474–499). New York: Van Nostrand Reinhold.

Roediger, H. L. (1991). They read an article? A commentary on the everyday memory controversy. *American Psychologist, 46,* 37–40.

Roediger, H. L., & Blaxton, T. A. (1987). Retrieval modes produce dissociations in memory for surface information. In D. S. Gorfein & R. R. Hoffman (Eds.), *Memory and cognitive processes: The Ebbinghaus Centennial Conference* (pp. 349–379). Hillsdale, NJ: Lawrence Erlbaum Associates.

Rumelhart, D. E., & McClelland, J. L. (1986). *Parallel distributed processing: Vol. 1. Foundations.* Cambridge, MA: MIT Press.

Rubin, D. C. (1989). Introduction to Part I: The how, when, and why of studying everyday cognition. In L. W. Poon, D. C. Rubin, & B. A. Wilson (Eds.), *Everyday cognition in adulthood and late life* (pp. 3–27). Cambridge, UK: Cambridge University Press.

Rybash, J. M., Hoyer, W. J., & Roodin, P. A. (1986). *Adult cognition and aging: Developmental changes in processing, knowing, and thinking.* New York: Pergamon.

Salthouse, T. A. (1991). *Theoretical perspectives on cognitive aging.* Hillsdale, NJ: Lawrence Erlbaum Associates.

Shiffrin, R. M., & Schneider, W. (1977). Controlled and automatic human information processing: II. Perceptual learning, automatic attending, and a general theory. *Psychological Review, 84,* 127–190.

Sinnott, J. D. (1989a). Background: About this book and the field of everyday problem solving. In J. D. Sinnott (Ed.), *Everyday problem solving: Theory and applications* (pp. 1–6). New York: Praeger.

Sinnott, J. D. (1989b). An overview—if not a taxonomy—of "everyday problems" used in research. In J. D. Sinnott (Eds.), *Everyday problem solving: Theory and applications* (pp. 40–54). New York: Praeger.

Sternberg, R. J. (1985). *Beyond IQ: A triarchic theory of human intelligence.* New York: Cambridge University Press.

Walters, J. M., & Gardner, H. E. (1986). The theory of multiple intelligences: Some issues and answers. In R. J. Sternberg & R. K. Wagner (Eds.), *Practical intelligence: Nature and origins of competence in the everyday world,*(pp. 163–182). New York: Cambridge University Press.

II TAXONOMY AND METHODOLOGY IN EVERYDAY COGNITION

2 What is Everyday Cognition?

Leonard W. Poon
Deborah J. Welke
William N. Dudley
University of Georgia

The debate on the best method used to examine cognitive mechanisms and behavior has been brought into focus in the last decade. Should behavioral studies be most efficiently conducted in the laboratory or in the natural environment? This chapter defines and illustrates this controversy as well as outlines our thinking in this area.

Our first section is designed to pinpoint and summarize the positions contained in two influential papers that most clearly represent the views of laboratory versus everyday cognitive research. Next, a survey of the literature from both domains is used to demonstrate similarities and differences of these approaches in relating to cognitive theories and applications. The third section of the chapter examines four common sources of confusion that tend to divide laboratory and ecological research. Finally, we attempt to answer the question, "*What is everyday cognition*?"

THE CONTROVERSY: LAB VERSUS LIFE RESEARCH

Two papers are available that we believe are most representative of the arguments in support of laboratory versus ecological approaches in cognitive research. On the one hand, Neisser (1978) presented an eloquent key note address at the 1978 Conference on Practical Aspects of Memory in which he argued for ecological research. On the other hand, Banaji and Crowder (1989) from Yale University provided an equally eloquent attack of the ecological research approach that was published in the *American Psychologist*. Both points of view are important to our understanding of everyday cognition. The following is a summary of their main arguments.

The title of Neisser's (1978) paper, "Memory: What are the important questions?", described his main criticism toward the traditional laboratory methods of memory research. He pointed out that laboratory based memory research has little to show for 100 years of effort for three general reasons. First, laboratory psychologists do not study interesting questions. "If X is an interesting or social significant aspect of memory, then psychologists have hardly ever studied X." (p. 4) Second, the reason laboratory-based researchers do not study interesting questions is that they feel they should study more fundamental questions that should lead to broader generalizations. However, Neisser charged that any grade school child already knows these broad generalizations about memory. Third, the reason traditional learning theory research collapsed is because its proponents failed to study ecologically relevant questions. Neisser argues that the modern information processing school will likely suffer the same demise for the same reason. He concluded that as naturalistic study of animal behavior has proven to be more rewarding than traditional research on learning, so a naturalistic study of memory may be more productive than its laboratory counterpart.

Steven Ceci (1990) provided a summary of the main arguments posed by Banaji and Crowder (1989) in a Psi Chi Dialogue devoted specifically to that paper at a meeting of the Southeastern Psychological Association. Ceci outlined five specific arguments posed by Banaji and Crowder in their paper, "The bankruptcy of everyday memory":

1. the goals of scientific study in natural and behavioral sciences are similar. Other sciences would have been hopelessly paralyzed in their effort to understand basic mechanisms if they had attempted to place equal emphasis on everyday applications.
2. laboratory controlled procedures are important to scientific methods. This level of control is frequently lacking in ecological research.
3. everyday memory research has yet to uncover new psychological principles.
4. the purpose of scientific methods is to expose invariant mechanisms whose detection should not be left entirely to naturalistic observation, retrospective analysis and the like.
5. a systematic knowledge is needed to separate the myths from the facts, and well controlled experimental techniques are needed to accomplish this goal.

In summary, these polemics have succeeded in soliciting discussion in support of either position and in exposing the strengths and weaknesses of these seemingly different approaches. To be sure, there are equally persuasive points and counterpoints on both sides of the argument. Just how disparate are these two approaches? In order to provide some qualitative insight, the next section of the paper examines the literature produced from the two perspectives in terms of their means (or methods), ends (or goals and contributions), as well as intervening constructs. The results from this analysis could provide an actual yard stick to evaluate the ongoing rhetorics and polemics.

A META-ANALYSIS

Literature Reviewed

Because a significantly greater amount of laboratory-based work has been reported in the literature compared to ecological-based studies, three domains of literature were sampled. First, we chose to compare the contents of two peer-reviewed journals: one that specializes in laboratory-based research and the other in ecologically-based behavioral research. *The Journal of Experimental Psychology: Learning, Memory and Cognition* (Vols. 1–4, 1987, 1988) was chosen as representative of the work in the laboratory research domain. *The Applied Cognitive Psychology* journal (Vols. 1–4, 1987, 1988) was selected to represent the ecological domain from the same time period. Finally, owing to the large amount of attention placed on the edited book, *Practical Aspects of Memory: Current Research and Issues,* (Gruneberg, Morris, & Sykes, 1988, Vol. 1) as an authoriative source for practical cognitive research, this volume was also selected to compare and contrast with the two peer-reviewed journals.

To increase the compatibility across the three sources, only empirically-based studies were used. No review articles were included. Further, only studies dealing with nonpathological, adult samples were examined.

About the same number of articles were selected from the three sources. Thirty-two papers were selected, according to a random number table, from the *Journal of Experimental Psychology* (JEP). Thirty-four papers were selected from the Gruneberg et al. volume and 30 papers from *Applied Cognitive Psychology (ACP).*

Questions Asked. In order to compare and contrast approaches to behavioral research from the laboratory and ecology perspectives, information on substantive questions, theory, methods, design, and conclusions were abstracted from the three literature domains.

1. Research question: We classified the primary purpose of a study on whether it asked a methodological or theoretical question, or examined a phenomenon. A study classified as *methodological* had a specific aim of determining the usefulness of a particular method under varying conditions, developing a new

method, or examining parametric manipulations. *Theoretical* studies compared or contrasted theories or attempted to contribute new data to a theoretical position. Studies in which the main purpose was to examine general, observable effects or behaviors but which did not test hypotheses, models, or theories were classified as *phenomenological.*

2. Rationale. Independent of the primary purpose of the study, we were interested whether the theoretical, methodological, or phenomenological concern was driven by, or justified by, a theoretical framework.

3. Research design. We classified research designs into the factorial, correlational, and survey/observational domains. A *factorial* design used clearly defined independent variables which were manipulated either within or between subjects. A *correlational* design described relationships between variables. A *survey/observational* study employed no active manipulation of variables by the experimenter and included survey studies, diary studies and naturalistic observation.

4. Stimuli/tasks. The level of realism and control of the stimuli and tasks employed was evaluated. A *laboratory-based* task employed a manipulation of stimuli or a procedure that an individual would not normally encounter in everyday life and in which experimental control of the procedure is easily applicable. Examples include Stroop color tests, mental rotation tasks, tachistiscopic displays, CVC tasks. A *realistic* task employed a manipulation of stimuli or a procedure that a person could encounter in everyday life. Examples include mailing postal letters, working math problems, viewing simulated events such as court room proceedings, and viewing actual events (traffic accidents) as an eyewitness.

5. Conclusions. The conclusion(s) stated by the author(s) not only could be used to confirm the specific aim of the paper, but would provide an indication of the directions in which the results could be implicated. In a *theory-based* conclusion the authors noted that the study either supported or failed to support the theoretical position that the study examined. In an *application-based* conclusion, the results were stated in terms of possible uses of the findings on a practical level. An *inconclusive* conclusion contained only a summary of the study or it contained no conclusion. A *method-based* conclusion emphasized the strengths or weakness of the particular method tested as compared to other methods. A *phenomenon-based* conclusion reiterated the contribution of the study to the understanding of a particular observed phenomena.

RESULTS

Figure 2.1 shows the frequency distribution of the type of research question posed in the three literature domains. As expected, theory-based research ques-

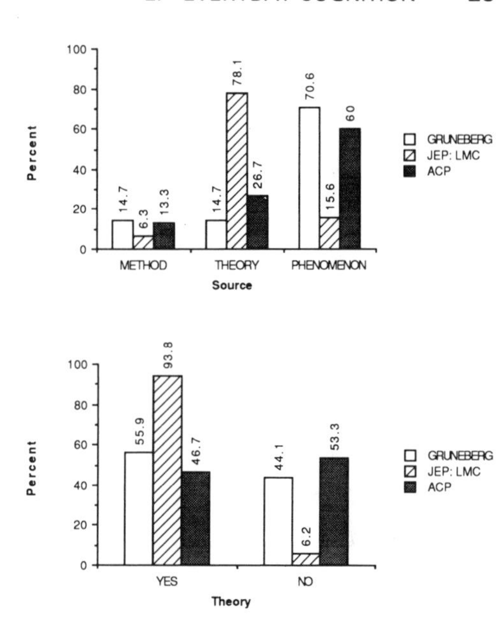

FIG. 2.1. Source of research question.

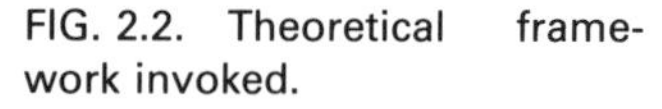

FIG. 2.2. Theoretical framework invoked.

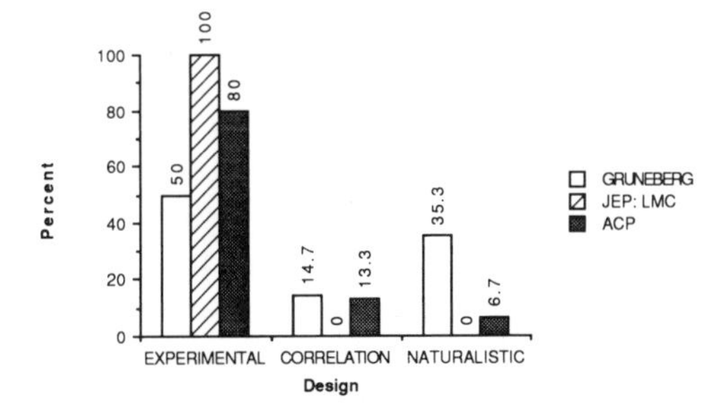

FIG. 2.3. Type of design.

tions predominated, 78%, in the *JEP* studies, while the other two applied sources examined phenomena-based questions, 60% to 70%. About a quarter of the *ACP* studies did examine theoretical issues or hypotheses. From 6% to 15% of the studies in the three domains examined methodological issues.

Figure 2.2 shows the distribution of theory and non-theory based studies. Again as expected, 94% of the *JEP* studies invoked a theoretical framework. It is interesting to note that about half of the applied literature sampled also invoked a theoretical framework.

Figure 2.3 shows the frequency of occurrence of three types of design in the samples. Studies reported in the two peer-reviewed journals used predominantly factorial designs, 100% in *JEP* and 80% in *ACP*, where design control could be easily imposed. In contrast, 50% of the studies contained in the non-peer review edited volume on practical behavioral research employed factorial designs; about

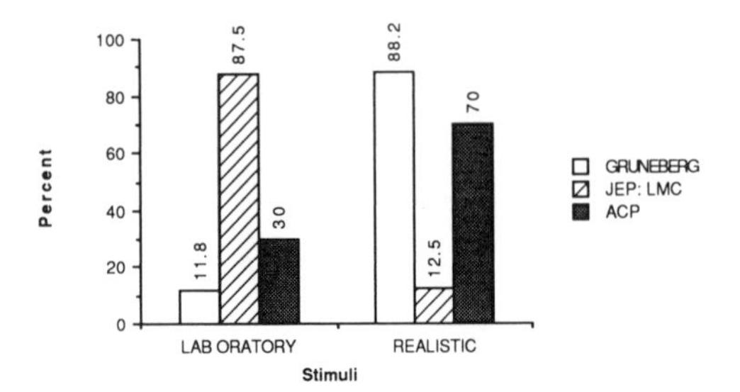

FIG. 2.4. Type of stimuli used.

35% of the reported studies also used naturalistic observation methods. *JEP* studies tended not to use correlational methods, while about 13% to 15% of the applied studies did.

Figure 2.4 shows the distribution of stimuli/tasks employed. The distribution confirms our expectation in that 87% of the *JEP* articles sampled used laboratory-based stimuli or tasks, which is consistent with a traditional experimental psychology approach. The Gruneberg et al. (1988) and *ACP* samples showed equally high percentages employing realistic stimuli or tasks, 88% and 70% respectively.

Figure 2.5 shows a distribution of the conclusions and implications drawn from the three domains of literature, and there are distinct differences. As expected, 87% of the *JEP* papers draw conclusions about their data within a theoretical framework in contrast to about 44% of the applied papers. In reflection of the quality of peer-reviewed papers, none of the *JEP* and only 3% of the *ACP* papers made inconclusive summaries; however, 18% of the edited chapters from Gruneberg et al. just related their findings. Two other observations should be noted when comparing the two applied literature sources: one, the Gruneberg chapters made a significantly greater number of conclusions based on applications (21% for Gruneberg and 7% for *ACP*), while the *ACP* papers made more conclusions about phenomena (37%) in comparison to the Gruneberg chapters (3%).

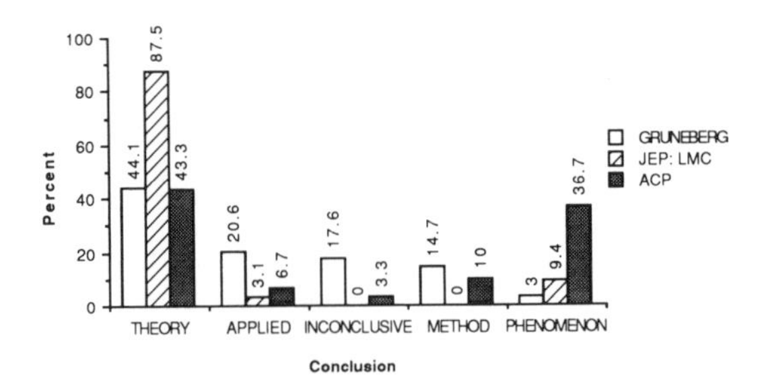

FIG. 2.5. Terms in which conclusions are stated.

WHERE'S THE BEEF?

Empirical analysis of studies from both laboratory-based and ecological-based domains shows that no clear monopoly exists concerning theory building, research design, and application of findings. That is, *JEP* and *ACP* studies are not entirely theoretical or applied in nature. Students in basic cognitive research should take comfort in the fact that studies reported in the *Journal of Experimental Psychology: Learning and Memory* are generally theory based and use factorial designs. Although stimuli used in *JEP* studies are primarily words, pictures, and artificial patterns, videotapes or simulation of everyday action were also employed. Many studies in both camps are phenomenon and theory driven with a number of studies also method driven or based entirely on parametric manipulations. Subtle differences exist between the peer-reviewed and non peer-reviewed ecological-based literature; however, by and large they tend to be similar. The important conclusion from this survey is that basic behavioral mechanisms can be uncovered and replicated in both laboratory and ecological settings (e.g., Baddeley, 1989; Bahrick, 1989; West, 1986).

Given that both laboratory and ecological studies can contribute to our understanding of cognition, we feel there are four commonly misunderstood issues that have contributed to the dichotomous viewpoints between proponents of laboratory and ecological research.

The *first* source of confusion is the tendency to dichotomize classes of stimuli and level of control used in laboratory and ecological research. Contrary to some polemic which would lead researchers to believe that tight control is possible only in a so-called laboratory situation, our meta-analysis of the procedures used in both camps showed that the selection of stimuli, subjects, and manipulation of dependent and independent variables all followed the same basic experimental psychology principles. In fact, it can be said that *both* camps employ laboratory-based research.

Perhaps the level of simulation of cognitive mechanisms under investigation may be a better classifier to describe experimental characteristics in the two camps. Although laboratory investigations are often driven by and generalized to everyday phenomenon (i.e., the use of paired associate or serial learning), there is no requirement that realistic stimuli be employed, although, as it can be seen in the survey of *JEP* studies, they can be. On the other hand, although ecological studies employ realistic stimuli to attempt to simulate actual conditions in everyday life, it is rare that these studies completely simulate all the everyday parameters (West, 1986). This issue of generalization from data to everyday life could be due to a minimal understanding of everyday cognitive mechanisms (by either camp) as well as the level of experimental control imposed that is necessary to make a study legitimate by existing criteria.

A *second* source of confusion in separating laboratory and ecological research

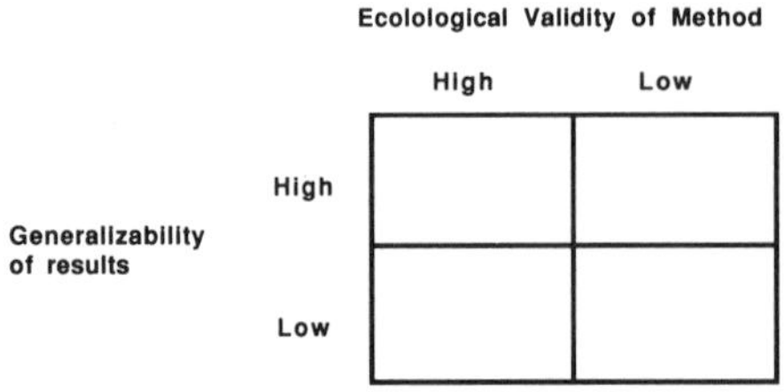

Ecolological Validity of Method

		High	Low
Generalizability of results	High		
	Low		

FIG. 2.6 Levels of generalization and ecological validity. From Banaji and Crowder (1989).

centers around issues of generalization, ecological and functional validity, and representativeness of design.

Banaji and Crowder (1989) argued that ecological validity of the methods is unimportant and can even work against generalization of findings. In a two by two array representing levels of generalization and ecological validity (Fig. 2.6), they presented cases to show that low ecological validity of methods and high generalization of results is preferred over high ecological validity and low generalization.

The implication is that the ecological nature of the situation or stimuli is of little effect or consequence as long as some basic principles of cognition can be discovered. It is likely that both proponents of laboratory and ecological research would agree with this conclusion. After all, this conclusion reflects basic experimental psychology axioms.

We would like to caution against oversimplication on the utility of ecological validity. Petrinovich (1989) painstakingly defined the Brunswikian model of ecological validity, which remained intact for more than 3 decades, but whose meaning has since eroded. Ecological validity refers to the degree to which variables relate to behavioral events; the quality of relationship between distal events and proximal stimulation is their ecological validity. Ecological validity is further differentiated from representativeness of design in the following manner. Ecological validity refers to the potential utility of various cues for the organisms in their environment, and representative design refers to the quality of naturalness, or lifelike quality, of the research. Petrinovich further pointed out that frequently when people speak in terms of ecological validity they often mean representativeness of the design (e.g., Bandura, 1978; Bronfenbrenner, 1977; Neisser, 1976).

Banaji and Crowder (1989) discussed their preference against lifelike situations. Their argument is more related to issues of representativeness of the design than to ecological validity. To the extent that utilization of cues in the environment is important for the manifestation of specific behavior, low ecologically valid cues may not be able to elicit the behavior. This situation can lead to erroneous conclusions and generalizations (e.g., Ceci & Bronfenbrenner, 1985) about basic cognitive mechanisms. The concept of ecological validity in the Brunswikian sense has its utility in understanding behavior and should not be dismissed.

Another class of distinction is also needed to clarify the confusion on ecological validity. While ecological validity refers to the structure of the environment, functional validity refers to the organism's use of that structure (Petrinovich, 1989). These two constructs could be orthogonal. For example, an organism may not use a stimulus (since it has a low functional validity) even though it is a good stimulus (has high ecological validity), or vice versa. To present a complete picture on generalizability of results, not only ecological but also functional validity must be considered.

A *third* source of confusion that could erroneously divide laboratory and ecological studies revolves around issues of control and generalization. A two by two matrix represents the different relationships between these variables (Fig. 2.7). Researchers are taught that control is desirable and is more easily obtained in the laboratory. Tighter control can lead to higher generalization to specific domains. Contrary to this perspective, there are a number of different opinions regarding control and generalization (Poon, Rubin, & Wilson, 1989). These viewpoints provide theoretical reasons for a choice between laboratory or ecological research designs (Rubin, 1989a). Two examples are provided here.

Rubin (1989b) argues for the demonstration of regularities and against control. He notes that given the present state of knowledge of cognitive mechanisms it is more important to establish and uncover regularities of phenomena than to exert experimental control. Once regularities are uncovered, then theories of cognition could be evoked or developed to understand that behavior. He argues that the naturalistic environment is the preferred laboratory since people display more regular behavior in less controlled situations. From this perspective, no control is preferred; lower control is related to higher regularity which in turn relates to higher genralizability.

Mook (1989) also advises against the traditional laboratory research method of exerting control in sampling subjects, stimuli, and situations. He reasoned that for most questions asked in research conducted outside the laboratory, control is neither needed nor desired. He has demonstrated by case studies (Mook, 1989) that published laboratory studies regularly violate every one of the canons of external validity in the research. For example, published examples could be found in which (a) assumptions of the representativeness of subjects, manipulations, and settings in relation to the research question were not met; (b) predic-

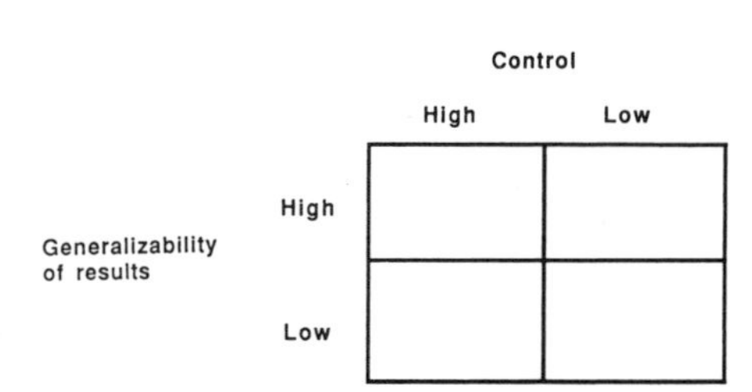

FIG. 2.7. Levels of generalization and control.

tion of real life from experimental findings is assessed but not easily demonstrable, and (c) the aim of accounting for a lot of real world variance is not met. He suggested that the notions of representativeness, prediction, and external validity must be dependent on the questions asked, not on some overarching rules of research.

Rubin (1989a), Mook (1989), and a number of other researchers (e.g., Baddeley, 1989; Bahrick, 1989; Bruce, 1989; Landauer, 1989; Petrinovich, 1989; and Sinnott, 1989) have presented alternative ways of examining behavioral mechanisms. These perspectives challenge the traditional ways researchers were taught about imposing control in experimental designs and in thinking about the generalization of results.

A *final* source of confusion between laboratory and ecological research could be clarified if the purpose of the research were articulated in terms of prediction and understanding. Again, we borrow from the work of Mook (1989) who identified how two models for research can provide different levels of information and generalization.

Figure 2.8 shows an "analogue" model of research as a guide to real-world prediction. Representative samples are drawn to generalize to a population. Agricultural research tends to fall within this category. Figure 2.9 shows an "analytic" model, which represents most of the work done by experimental and cognitive psychologists. The research begins and ends with a substantive question. In this model, there is no "population" so that the notion of representative-

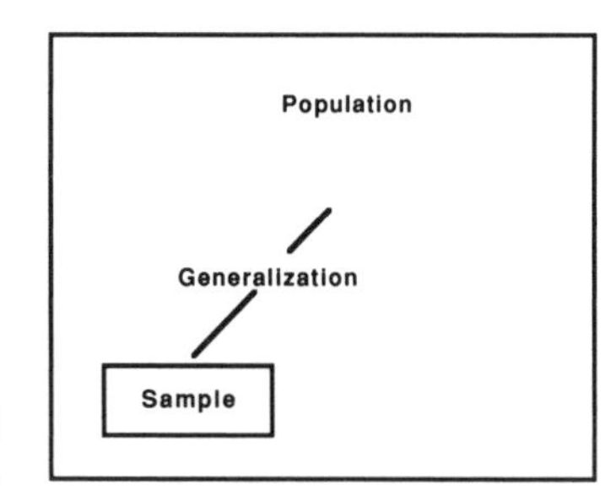

FIG. 2.8. The analogue model of research. From Mook (1989).

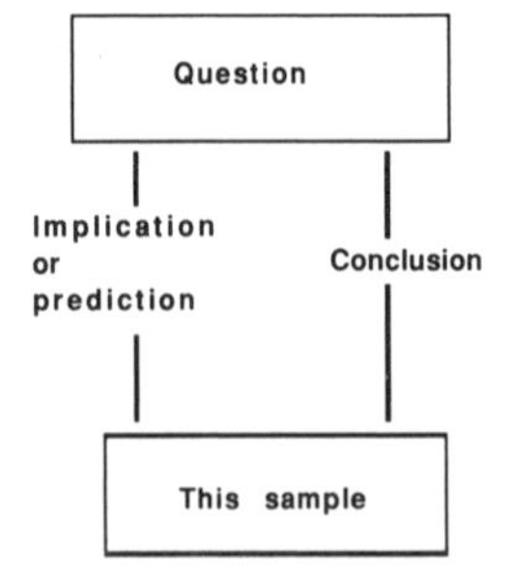

FIG. 2.9. The analytic model of research. From Mook (1989).

TABLE 2.1
Models and Settings

Model	*Laboratory*	*Natural Environment*
Analogue	A. Laboratory analogues or models of real-life phenomena designed to predict events in, or to estimate the characteristics of, a population of natural events.	B. Field observations, experiments, or correlational studies designed to predict or estimate events or characteristics of apopulationof such events.
Analytic	C. Laboratory investigations of the functions or mechanisms of mind or behavior.	D. Field observations, experiments, or correlational studies designed to elucidate the functions or mechanisms of mind or behavior.

From Mook (1989)

ness does not apply. Generalization is often not necessary since the sample is not measured to estimate what the general case is like. Rather the researcher asks: How do these subjects, in this setting, respond to this manipulation?

In Table 2.1 we see how the analogue and analytic models of research fit into laboratory and naturalistic environments. Medical research would fall into cell A. Survey and marketing research would fall into cell B. Experimental psychology research would belong to cell C, and psychological field research and correlational studies would belong to cell D. Mook helps us to make decisions on where and how to conduct our studies by the questions and outcomes we seek. In this manner the controversy and confusion regarding the purpose of conducting experiments in the laboratory and ecological setting can be made moot by Mook.

WHAT IS EVERYDAY COGNITION?

We began to answer this question by introducing our own view of the current thinking about everyday cognition research. Then we reviewed the pertinent papers representing opposing views and performed a meta-analysis to assess similarities and differences. Finally, we examined four specific issues that tend to confuse and divide research in the laboratory and ecological settings.

We concluded that laboratory-based and ecologically based research are two faces of the same coin. The coin represents the cumulation of systematic knowledge about cognitive mechanisms. We acknowledge that the debate on the relative efficiency of knowledge accumulation is real, but also that the debate represents the age-old tension between the need for control and the need for the preservation of the essence of the phenomenon under investigation.

We are not so naive to think that our rational examination of some controversial issues could change entrenched opinions about the perils of research. On the contrary, we appeal to our peers to consider different strategies and experiments that could ultimately unearth the secrets of cognitive mechanisms. After all, this is the basic aim of all cognitive scientists. The motivation to understand invariant cognitive mechanisms and a family of mechanisms would necessarily move researchers to sample behavior, situations, and subject groups using both laboratory and ecological methods.

The important questions are not *whether* studies should be performed in the laboratory or the naturalistic environment, but rather *when* the condition is most favorable to perform an experiment in a laboratory or an ecologically valid setting, and *why* a particular setting should be selected (Landauer, 1989; Rubin, 1989a). If the study of cognitive mechanisms is approached from this perspective, then there is no need to artificially divide laboratory and ecological research.

The following are some suggestions based on our analyses and other sources (e.g., Poon et al., 1989) for making decisions on the appropriate setting and experimental design in the quest to understand cognitive mechanisms:

1. An individual's behavioral and cognitive manifestations are dependent on situational, biographical, and historical contexts. The substantive research question about cognitive mechanisms dictates the relative scope of the contextual involvement and ultimately the experimental design in the study of a phenomenon.
2. The level of understanding regarding a family of cognitive mechanisms must take into account these contextual variables.
3. Regularity of a phenomenon within a context provides one indication of the relative meaningfulness of the phenomenon.
4. The level of control imposed on the selection of stimuli, environment, and procedure in an experiment would affect the generalizability of results and regularity of patterns of behavior.
5. Generalizability of results depends not only on the similarity of the stimuli to the actual phenomenon (representativenss of the design) but the proximal relationship to behavioral events and the utilization of functional cues within that context (ecological and functional validity).
6. The selection of the naturalness of a research setting depends on whether the research goal is to predict, or understand, or both.

SUMMARY AND CONCLUSION

We would like to make two points to conclude our chapter. First, natural or behavioral phenomena never operate under conditions of complete closure (Manicus & Secord, 1983). This explains why behaviors and experiences are described by patterns, tendencies, and probabilities. Second, events are always the outcome of complex causal configurations. This explains why cognitive phenomena are investigated with varying levels of control, stimuli, tasks, subjects, and environments.

Finally, specific behavior cannot be explained as the simple manifestation of a single law or theory. The explanation of behavior is properly a multidisciplinary effort necessarily transcending biology and social sciences and should include the measurement of behavior in and out of the laboratory.

REFERENCES

Baddeley, A. D. (1989). Finding the bloody horse. In L. W. Poon, D. C. Rubin, & B. A. Wilson (Eds.), *Everyday cognition in adulthood and late life* (pp. 104–115). Cambridge, UK: Cambridge University Press.

Bahrick, H. P. (1989). The laboratory and ecology: Supplementary sources of data for memory research. In L. W. Poon, D. C. Rubin, & B. A. Wilson (Eds.), *Everyday cognition in adulthood and late life* (pp. 73–83). Cambridge, UK: Cambridge University Press.

Banaji, M. R., & Crowder, R. G. (1989). The bankruptcy of everyday memory. *American Psychologist, 44,* 1185–1193.

Bandura, A. (1978). On paradigms and recycled ideologies. *Cognitive Therapy and Research, 2,* 79–103.

Bronfenbrenner, U. (1977). Toward and experimental ecology of human development. *American Psychologist, 32,* 513–531.

Bruce, D. (1989). Functional explanations of memory. In L. W. Poon, D. C. Rubin, & B. A. Wilson (Eds.), *Everyday cognition in adulthood and late life* (pp. 44–58). Cambridge, UK: Cambridge University Press.

Ceci, S. J. (1990, April). Are everyday cognition and memory solvent? In L. W. Poon (Chair), Panel discussion conducted at the meeting of the Southeastern Psychological Association, Atlanta, Georgia.

Ceci, S. J., & Bronfenbrenner, U. (1985). "Don't forget to take the cupcakes out of the oven": Prospective memory, strategic time-monitoring, & context. *Child Development, 56,* 152–164.

Davies, G., & Hermann, D. J. (Eds.). (1987). *Applied Cognitive Psychology, 1*(1–4).

Davies, G., & Hermann, D. J. (Eds.). (1988). *Applied Cognitive Psychology, 2*(1–4).

Gruneberg, M. M., Morris, P. E., & Sykes, R. N. (Eds.) (1988). *Practical aspects of memory: Current research and issues. Vol. 1.* New York: Wiley.

Landauer, T. K. (1989). Some bad and some good reasons for studying memory and cognition in the wild. In L. W. Poon, D. C. Rubin, & B. A. Wilson (Eds.), *Everyday cognition in adulthood and late life* (pp. 116–125). Cambridge, UK: Cambridge University Press.

Manicus, P. T., & Secord, P. F. (1983). Implication for psychology of the new philosophy of science. *American Psychologist, 38,* 399–413.

Mook, D. G. (1989). The myth of external validity. In L. W. Poon, D. C. Rubin, & B. A. Wilson (Eds.), *Everyday cognition in adulthood and late life* (pp. 25–43). Cambridge, UK: Cambridge University Press.

Neisser, U. (1976). *Cognition and reality.* San Francisco: W. H. Freeman.

Neisser, U. (1978). Memory: What are the important questions? In M. M. Gruneberg, P. E. Morris, & R. N. Sykes (Eds.), *Practical aspects of memory* (pp. 3–24). London: Academic Press.

Petrinovich, L. (1989). Representative design and the quality of generalization. In L. W. Poon, D. C. Rubin, & B. A. Wilson (Eds.), *Everyday cognition in adulthood and late life* (pp. 11–24). Cambridge, UK: Cambridge University Press.

Poon, L. W., Rubin, D. C., & Wilson, B. A. (Eds.). (1989). *Everyday cognition in adulthood and late life*. Cambridge, UK: Cambridge University Press.
Roediger III, H. L., Medin, D. L., & Smith, M. C. (Eds.). (1987). *Journal of Experimental Psychology: Learning, Memory and Cognition, 13*(1–4).
Roediger III, H. L., Medin, D. L., & Smith, M. C. (Eds.). (1988). *Journal of Experimental Psychology: Learning, Memory and Cognition, 14*(1–4).
Rubin, D. C. (1989a). Introduction to Part I: The how, when, and why of studying everyday cognition. In L. W. Poon, D. C. Rubin, & B. A. Wilson (Eds.), *Everyday cognition in adulthood and late life* (pp. 3–7). Cambridge, UK: Cambridge University Press.
Rubin, D. C. (1989b). Issues of regularity and control: Confessions of a regularity freak. In L. W. Poon, D. C. Rubin, & B. A. Wilson (Eds.), *Everyday cognition in adulthood and late life* (pp. 84–103). Cambridge, UK: Cambridge University Press.
Sinnott, J. D. (1989). General systems theory: A rationale for the study of everyday memory. In L. W. Poon, D. C. Rubin, & B. A. Wilson (Eds.), *Everyday cognition in adulthood and late life* (pp. 59–70). Cambridge, UK: Cambridge University Press.
West, R. L. (1986). Everyday memory and aging. *Developmental Neuropsychology, 2*(4), 323–344.

3 Everyday Cognition: Taxonomic and Methodological Considerations

Sherry L. Willis
K. Warner Schaie
The Pennsylvania State University

Several years ago one of our research subjects asked a question that has influenced our present work and is central to the topic of this volume. Our research has focused broadly on a psychometric abilities approach to the study of adult intelligence. Since intelligence is construed in many different ways by lay persons and particularly by older adults, the term "problem solving" has often been used in explaining our research to subjects. One day as one of our staff members was explaining a study, an elderly farm woman interrupted and asked: "Now, dear, are we going to solve *your* problem or *my* problem?" There is the assumption in research on everyday cognition that the problems that our subjects encounter in their daily lives are the focus of study. In this chapter we consider not only research findings and methodological issues, but also examine *whose* problem is really being studied.

DEFINING EVERYDAY COGNITION

A major focus of this chapter is on methodological issues related to the study of everyday cognition. However, the phenomenon must be delineated before considering methods and procedures for studying it. Although there is no commonly agreed upon definition for everyday cognition, there is some consensus regarding certain dimensions characterizing the phenomenon. Several definitions of everyday cognition are presented in Table 3.1. Three dimensions are highlighted in these definitions (Charlesworth, 1976; Neisser, 1976; Wagner, 1986). First, everyday problem solving involves the application of *cognitive abilities and skills*. Second, practical problems are experienced in *naturalistic or everyday contexts*.

TABLE 3.1
Definitions of Everyday Cognition

INTELLECTUAL COMPETENCE in NATURALISTIC settings or in wordly affairs (Neisser, 1976; Wagner, 1986)
Behavior under control of COGNITIVE PROCESSES and employed toward the solution of problems which challenge the WELLBEING, NEEDS, PLANS, and SURVIVAL of individuals (Charlesworth, 1976)
Practical Intelligence: (a) Is embedded in the individual's ORDINARY EXPERIENCE; (b) Is most often formulated by the individual and is of INTRINSIC INTEREST to the problem solver; (c) May be ILL DEFINED or involve ill structured information; (d) Involves MULTIPLE ANSWERS or MULTIPLE METHODS of solution (Neisser, 1976; Wagner & Sternberg, 1985).

Third, everyday problems are *complex and multidimensional,* when compared to laboratory tasks, such as simple reaction time or list learning.

Some authors have added other characteristics on which there is far less consensus (Neisser, 1976; Sternberg & Wagner, 1986; Sinnott, 1989). These include statements asserting that everyday problems are of necessity ill structured, are defined primarily or solely by the problem solver, and must involve alternative solutions. We question whether these latter attributes necessarily characterize many of the problems encountered by individuals in their everyday lives. There are many everyday tasks that are fomulated by society rather than by the individual (e.g., driving regulations, banking procedures). Also, many everyday tasks involve one correct or commonly agreed upon solution or procedure (e.g., determining the correct departure time from an airline schedule; giving the correct currency in making a purchase).

Just as there is little concensus on the definition of everyday cognition, there is also little commonality across investigators in the types of problems or tasks employed to study everyday cognition. One approach has been the study of expertise—comparing experts and novices on the specialized skills and knowledge bases associated with a profession or field. Skills associated with complex tasks, such as race course handicapping, chess, mental squaring, typing, and interpretation of medical slides have been studied (Ceci & Liker, 1986; Charness, 1989; Hoyer, 1985; Salthouse, 1990). While mastery of these types of skills is critical in certain professions, these are not representative of the problems experienced by most adults in their daily lives.

The Instrumental Activities of Daily Living (IADLs) is a more common universe of daily activities (Fillenbaum, 1985; Lawton & Brody, 1969). Ability to accomplish these activities is considered essential in order to live independently in our society. Our most recent research has focused on the cognitive demands of seven IADL domains (Willis, 1991). These domains are shown in the first column of Table 3.2. Older adults' comprehension of printed materials (directions, charts, forms) related to each of the IADL domains is being examined. Exemplar tasks are shown in the second column of the table.

TABLE 3.2
Instrumental Activities of Daily Living (IADLs)

IADL Domains	*Exemplar Task*
TELEPHONE USAGE	Dialing instructions
SHOPPING	Directions for coupon redemption
TRANSPORTATION	Bus schedule
HOUSEKEEPING	Washing machine instructions
FOOD PREPARATION	Nutrition label
MEDICATIONS	Medicine bottle label
FINANCIAL MANAGEMENT	Billing statement

EVERYDAY COGNITION: TAXONOMIC ISSUES

We have conceptualized everyday problem solving as involving multiple dimensions as shown in Table 3.3. The cognitive core of everyday problem solving involves the assumption of the presence of a sufficient level of relevant mental abilities and domain-specific knowledge bases required to solve the problem (Park, 1991; Willis, 1991; Willis & Marsiske, 1991). The knowledge base and abilities are then integrated with certain social and affective components of everyday problem solving. Everyday problem solving is a highly individualized affair, and thus subjects' understanding and perceptions of their own personal circumstances enter into the process. The individual must consider a given problem within a specific physical and interpersonal context (Lawton, 1982). One's attitudes and beliefs, such as locus of control and self-efficacy play a role in considering solution alternatives (Baltes & Baltes, 1986; Rodin, Timko, & Harris, 1985). Finally, resolution of a problem typically requires the integration of all of the preceding dimensions or phases. This integration process involves: Awareness of the problem, the identification of alternative options for solving the problem, the ruling out of options that will not work, and the prioritizing of viable options. Selecting the "best" option should involve the preceeding dimensions, including domain-specific knowledge base, personal circumstances, as well as attitudes, beliefs, and preferences.

TABLE 3.3
Dimensions of Everyday Cognition

A. Relevant abilities and skills
B. Domain-specific knowledge bases
C. Understanding of:
    Personal circumstances
    Interpersonal context
D. Attitudes, beliefs, preferences
E. Integration of dimensions

The Role of Cognition in Everyday Problem Solving

We now consider in more detail cognition in relation to everyday problem solving, since much of our research has been focused at this level. Three questions are of interest: (a) What are the cognitive demands of the problem to be solved? (b) What is the relationship between the cognitive demands of a specific problem and more traditional mental abilities and information processing approaches to the study of cognition? (c) What is the role of domain-specific knowledge and everyday cognition?

Cognitive Demands of Everyday Tasks. Excellent work on analyzing the cognitive demands of tasks is being conducted by those studying expertise. Their research has focused on modeling the processing demands of complex tasks such as race course handicapping (Ceci & Liker, 1986), bridge and chess (Charness, 1981, 1983), typing (Salthouse, 1990), and the interpretation of medical slides (Hoyer, 1985).

In our research on everyday problems involving printed materials, we have adapted Meyer's parsing scheme for application with nontextual materials, such as charts and forms (Meyer, Marsiske, & Willis, 1993). Utilizing the parsing scheme, several components related to the difficulty of comprehending the printed material have been identified. Components associated with the difficulty level of the printed material include factors such as the length of the text, the level in the text structure of the salient information, whether the salient information is signaled (e.g., capitalized or italicized) in the material, and the number of points in the text structure that must be searched in order to answer the question (Meyer et al., 1993). Estimates of the difficulty level of a problem can be derived or the relative difficulty of two different problems can be compared. In Fig. 3.1, for example, two stimuli, dialing instructions for a pay phone and a chart of alternative medigap insurance plans, are presented from our battery of everyday printed materials. The relative difficulty of answering a question with respect to each of the stimulus materials was analyzed utilizing the parsing scheme and the components of difficulty. Figure 3.2 presents the relative difficulty level of the phone and insurance problems for several of the components derived from the text analysis research.

There are a number of limitations for both the expertise and text analysis approaches to analyzing the cognitive demands of everyday problems. First, the analyses are very labor intensive. Second, the modeling and processing components derived are often highly task specific. The modeling of the cognitive demands of even the most important tasks of daily living looms as a herculean endeavor. In studying the cognitive demands of a broad array of everyday problems, task taxonomies will need to be developed so that it becomes possible to compare different tasks along common dimensions.

Everyday Problem Solving and Psychometric Intelligence: What is the Relationship? A second question deals with the relationship between the cognitive demands of everyday tasks and the mental abilities and processes that have traditionally been investigated in the study of adult intelligence (Schaie, 1983). Recently, it has been proposed that intelligent behavior involves multiple forms of intelligence (Baltes, Dittman-Kohli, & Dixon, 1984; Sternberg & Berg, 1987). The distinction has been made between the "mechanics of intelligence," involving basic mental abilities and cognitive processes, and the "pragmatics of intelligence" concerned with everyday cognition. A major issue for multiple intelligence theories is the nature of the interrelationship among various forms of intelligence.

A mechanistic theory is needed to identify the cognitive processes involved in practical problem solving. However, the question arises whether the mental abilities and processes traditionally studied by psychologists are even relevant to the study of everyday cognition. Some contend that practical intelligence and more traditional approaches (e.g., psychometric intelligence) are distinct and unrelated forms of intelligence (Ceci & Liker, 1986; Friedricksen, 1986). Others, including ourselves, find a hierarchical relationship more plausible. From a hierarchical perspective, basic cognitive processes and abilities are believed to be universal across cultures and contexts. When nurtured and directed by a particular context, cognitive processes and abilities develop into domain-related competencies, that are manifested in daily life as cognitive performance. This view is supported by several recent studies that have found significant relationships between everyday task performance and traditional intelligence measures (Willis, 1991). Fluid and crystallized intellectual abilities have been found to be related to everyday problem solving (Camp, Doherty, Moody-Thomas, & Denney, 1989), interpersonal competence (Cornelius & Caspi, 1987), computer literacy (Garfein, Schaie, & Willis, 1988), and comprehension of printed materials (Willis & Schaie, 1986; Willis & Marsiske, 1991; Willis, Jay, Diehl & Marsiske, 1992). Capon and Kuhn (1979, 1982) reported consumer behavior to be related to formal reasoning within Piagetian theory. Proficiency in leisure activities, including recall of TV shows, video games, jigsaw, and crossword puzzles, were found to be predicted by verbal ability (Cavanaugh, 1983; Willis, Maier, & Tosti-Vasey, 1992), reaction time (Clark, Lanphear, & Riddick, 1987), and memory (Rice, Meyer, & Miller, 1988).

A Hierarchical Perspective. Several assumptions need to be explicated regarding a hierarchical relationship between abilities and everyday tasks. First, because practical intelligence tasks are complex, everyday problem solving involves the application of *multiple* mental abilities and cognitive processes (Willis & Marsiske, 1991; Willis & Schaie, 1986). For example, our research indicates that comparing alternative health insurance plans, like the one shown in Fig. 3.1,

Chart: Group Health Insurance Program

MEDICARE SUPPLEMENT PLANS

MEDICARE SUPPLEMENT INSURANCE PLANS	HOSPITAL BENEFITS	COVERS DOCOTOR AND MEDICAL CARE IN AND OUT OF HOSPITAL (AFTER $75 PLAN DEDUCTIBLE)*		SKILLED NURSING FACILITY BENEFITS		IN-HOSPITAL PRIVATE DUTY NURSING		PRESRIPTION DRUGS
	INPATIENT HOSPITAL DEDUCTIBLE COVERAGE	20% OF MEDI-CARE ELIGIBLE EXPENSES NOT PAID BY MEDICARE	COSTS ABOVE MEDICARE ELIGIBLE EXPENSES**	DAYS 1-18	DAYS 151-365	UP TO $30 PER SHIFT	80% OF USUAL AND PREVAILING CHARGES***	
MEDICARE SUPPLEMENT		✓		✓	✓	✓		
MEDICARE SUPPLEMENT PLUS	✓	✓		✓	✓	✓		
EXTENDED MEDICARE SUPPLEMENT	✓	✓		✓	✓		✓	✓
COMPREHENSIVE MEDICARE SUPPLEMENT	✓	✓	✓	✓	✓		✓	✓

Directions: Pay Phone Instructions

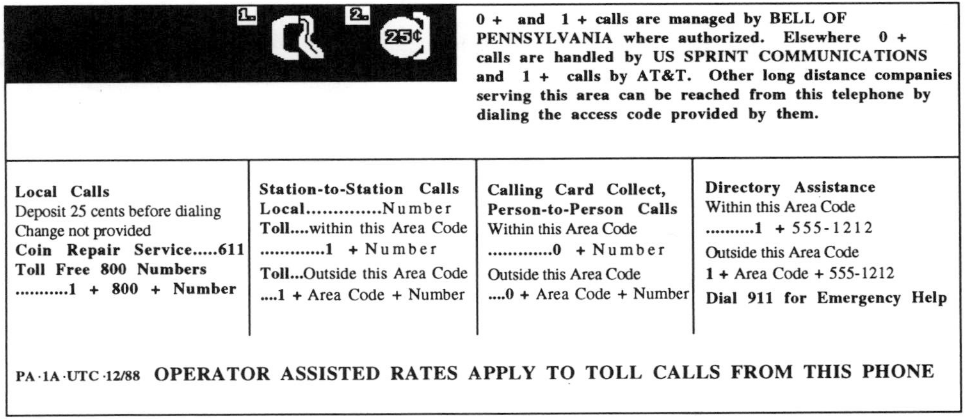

1. 2. 25¢

0 + and 1 + calls are managed by BELL OF PENNSYLVANIA where authorized. Elsewhere 0 + calls are handled by US SPRINT COMMUNICATIONS and 1 + calls by AT&T. Other long distance companies serving this area can be reached from this telephone by dialing the access code provided by them.

Local Calls	**Station-to-Station Calls**	**Calling Card Collect, Person-to-Person Calls**	**Directory Assistance**
Deposit 25 cents before dialing Change not provided **Coin Repair Service.....611** **Toll Free 800 Numbers** **..........1 + 800 + Number**	**Local.............**Number **Toll....**within this Area Code **............1** + Number **Toll...**Outside this Area Code **....1** + Area Code + Number	Within this Area Code **............0** + Number Outside this Area Code **....0** + Area Code + Number	Within this Area Code **.........1** + 555-1212 Outside this Area Code **1** + Area Code + 555-1212 **Dial 911 for Emergency Help**

PA·1A·UTC·12/88 **OPERATOR ASSISTED RATES APPLY TO TOLL CALLS FROM THIS PHONE**

FIG. 3.1. Stimulus material for everyday tasks measure. Top figure is chart of group health insurance plans. Bottom figure is directions for making long distance calls from pay phone.

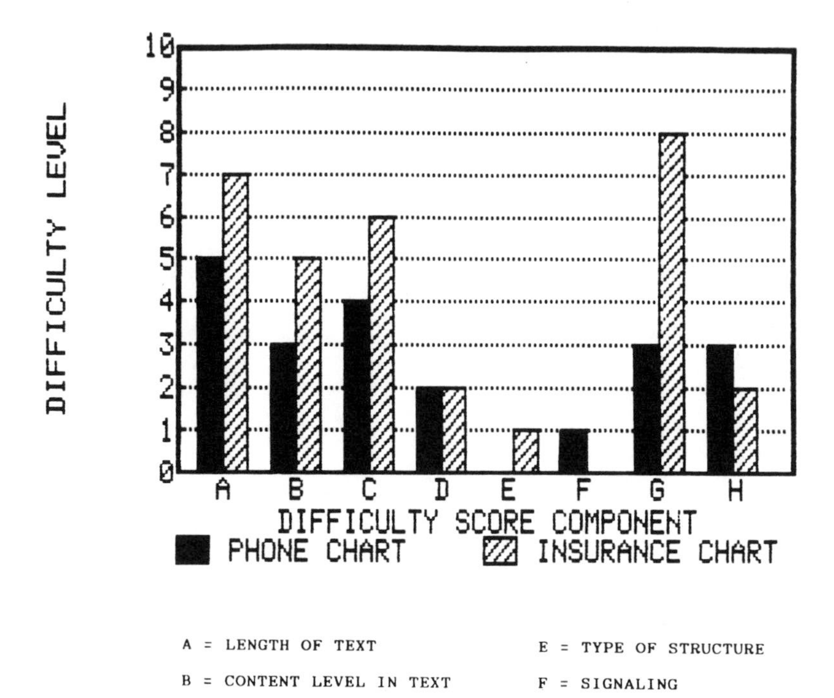

FIG. 3.2. Comparison of difficulty level of problems involving insurance plan and phone instructions, based on readability components.

involves both verbal and inductive reasoning ability. While verbal ability is required to read the chart, making comparisons among different insurance plans involves inductive reasoning.

Second, *different* constellations of mental abilities and processes will be required for various types of practical problems (Willis, 1991; Willis et al., 1992). For example, spatial orientation ability will be more salient for reading a map, whereas inductive reasoning will be more salient for interpreting a drug label. Third, competence with respect to basic abilities and processes is seen as a *necessary but not sufficient condition* for the successful solution of practical problems. Affective and social components are also involved, and are briefly considered later in this chapter.

Domain-Specific Knowledge

The assumption is often made that effective everyday problem solving is strongly related to the breadth and depth of an individual's domain-specific knowledge base. However, our current understanding of the importance of domain-specific knowledge bases for the solution of applied problems is based largely on the expertise literature. Here, indeed, domain-specific knowledge bases have been shown to play a critical role (Hoyer, 1985; Salthouse, 1984, 1990). Whereas the expertise literature has shown relationships between a single, highly specialized knowledge base and competence in a skill or profession, it is likely that solutions for many everyday types of tasks will require accessing *several* different knowledge bases. Again, referring to our insurance chart (Fig. 3.1), the problem solver would need to have at least rudimentary knowledge not only about insurance policies but also about the medicare system. Mapping out the particular set of knowledge bases required for a given practical problem remains an important task for future research.

Even more problematic, however, is the lack of measures or procedures for assessing an individual's breadth or depth of knowledge in many substantive domains. One approach has been to use subjects' self-reports of frequency of performing a task or their familiarity with a task as proxies for their level of domain-specific knowledge and skill (Salthouse, 1990). It has been assumed that the greater amount of time spent in a task will always result in an increase in domain-specific knowledge and skill. However, empirical findings have been quite mixed on the relationship between the frequency of carrying out everyday tasks and competence in task performance (Cornelius & Caspi, 1987; Rice et al., 1988). What accounts for this discrepancy? Learning of complex skills generally involves the application of these skills to a diverse set of problems and in a wide array of contexts, so that general principles can be developed (Schooler, 1989; Sprafkin & Goldstein, 1990). But many older persons do not experience this diversity of problems and contexts in their daily lives. Although they may spend considerable time on a particular task, their experience is often confined to the same context and exercised in a routinized and repetitive manner. In assessing task familiarity, we must therefore go beyond studying the length of time spent on a given task to examining *how* time was spent in task-related activities.

Knowledge of Personal Circumstances and Interpersonal Context

Considered next are the more personal, affective, and social dimensions of everyday problem solving in which individuals take into account their own personal circumstances and contexts in solving tasks of daily living. Consider again the older adult faced with the problem of purchasing a supplemental medigap health insurance plan. Which insurance plans represent viable options for an elderly woman, for example, will be partially determined by personal circumstances: her financial status, as well as her current and future health status. Likewise, the interpersonal context of the individual must be taken into account. For example, what types of health care services need to be purchased will depend in part on the older adult's social support network. The availability of family and friends to provide assistance must be factored into decisions regarding health care needs (Moss & Lawton, 1982). The availability of alternative health care services in the physical environment must also be considered.

Attitudes, Beliefs, and Preferences

Understanding of one's personal circumstances reflects in part certain attitudes, beliefs, and preferences (Baltes & Baltes, 1986). For example, it is likely that health-related locus of control and self-efficacy beliefs (Wallston & Wallston, 1982) will influence one's decisions regarding health insurance. Current research on age-related changes in self-efficacy indicates an increased dependence on powerful others in old age (Lachman, 1986; Rodin et al., 1985). Some elderly persons may therefore increasingly seek and depend on the advice of others in making important decisions in everyday life.

Integration of Everyday Cognition Dimensions

Effective solution of everyday problems requires, of course, an integration of the various dimensions discussed earlier (Table 3.3). Integration is ongoing, occuring at various phases of the problem solving process, as well as at the point of reaching a final decision or solution. Effective integration involves several steps: (a) Identification of solution alternatives; (b) ruling out of options that will not work given the individual's personal circumstances; and (c) prioritizing the remaining viable options.

In decision theory, the first step involves the ability to consider a problem from multiple perspectives and to identify alternative solutions. This step has been emphasized in some research. These studies have judged effective everyday problem solving in terms of how many alternative solutions the subject considered (Camp et al., 1989; Denney, 1989). However, while there may be a large number of hypothetical solution alternatives, in many everyday problem solving situations most adults are faced with painfully few real options. For example, although most medigap insurance plans offer four to six options, most older adults indicate that not more than two options are really viable, given their personal circumstances. Thus, in practical problem solving, the ruling out of options that are not feasible or will not work, and the prioritizing of the remaining viable options may be more crucial than the generation of a large number of hypothetical alternative solutions.

Problem Solving as a Recursive Process

Our laboratory experiments have left us with the naive assumption that problem solving is linear and is consummated relatively quickly. However, problem solving in the real world is often recursive (see Fig. 3.3) and extends over days, months, or years. The individual may exercise through the dimensions of problem solving multiple times before reaching a final solution. Alternatively, a problem may appear to be "solved" at one point in time only to recur, requiring it

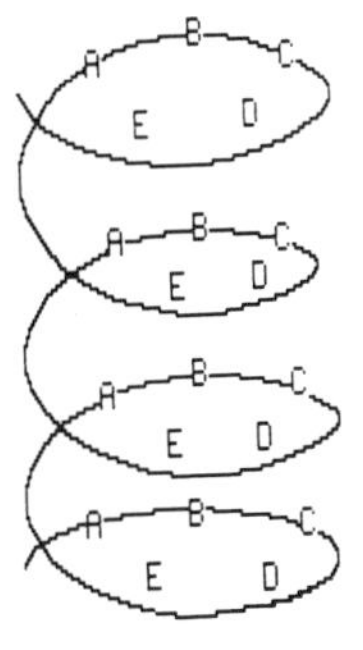

FIG. 3.3. Everyday problem solving is a recursive process. At each cycle in the recursive process, any or all of the dimensions in problem solving (Table 3.3) may have changed.

to be resolved at a later time. At each cycle in the recursive process, any or all of the dimensions (see Table 3.3) in problem solving may have changed. For example, the problem of choosing among medigap insurance options must be repeatedly resolved as medicare benefits shift and as the individual's health care needs change. Within each cycle, there may also be shifts in the individual's cognitive competence to consider the options as well as in the person's self-efficacy beliefs regarding their health.

METHODOLOGICAL ISSUES

As our delineation of the dimensions of problem solving suggests, the process by which an individual solves everyday problems may not be entirely rational. However, it does not follow that the methods and procedures for the study of everyday cognition must also be ill structured and ambiguous. Instead, we argue that the "murkier" the problem, the more rigorous must be the approach to the study of that problem. Because neither the domain of everyday problem solving nor the procedures for its study are well defined, structure must be imposed by developing taxonomies for the study of everyday problem solving. In this section four questions will be addressed. First, there is problem definition—that is, what are the *elements* currently being employed in studying the domain of everyday problem solving? Second, there are issues regarding *subject sampling*. Third, the *methods of assessing* everyday cognition that are currently employed are examined. Finally, the thorny question of how we are to judge the *adequacy of problem solutions* is addressed.

Dimensionalizing the Domain of Everyday Problem Solving

A cursory review of the practical intelligence literature suggests that the study of everyday problem solving has followed three approaches, each of which has addressed a different facet of the phenomenon. These facets have been labeled: Personal attributes, Task properties, and Contexts (see Fig. 3.4).

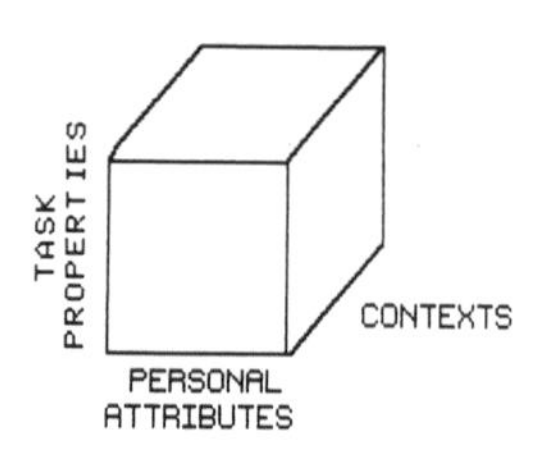

FIG. 3.4. Three facets of everyday problems: Personal attributes, task properties, and context. Study of everyday problems involves examination of the interrelationship among these facets.

Personal Attributes. In the first approach, the focus is on the problem solver. What are the personal attributes, characteristics, skills, or tacit knowledge exhibited by an intelligent, competent adult? In some studies, there has been a general search for attributes and characteristics of an intelligent person, one who possesses whatever it is that might be characterized as practical intelligence. In this type of study there is no reference to solving a particular problem. This approach is reflected in studies concerned with naive theories of intelligence (Berg, 1990). In his research on implicit theories of intelligence, Sternberg (e.g., Sternberg, Conway, Keton, & Berstein, 1981) compared attributes of an intelligent person, as defined by lay persons vs. academic psychologists.

The more common approach has been to examine attributes and skills of the individual with respect to solving a particular type of problem. In studies focusing on common, everyday problems, the problem solver's level of intellectual ability is related to problem solving proficiency (Camp et al., 1989; Willis, 1991; Willis et al., 1992). In studies of expertise, domain-specific skills and knowledge are related to level of proficiency (Charness, 1989; Salthouse, 1987).

Various procedures have been employed to identify or define the attributes or competences of the problem solver. One procedure employs the ratings of judges, as in defining an individual's level of mastery in chess or bridge. Alternatively, experimental procedures have been employed to differentiate the level of skill and cognitive processes employed by experts and novices, such as in Salthouse's study (1984) of expert older typists. In ethnographic research, interviews and observational procedures have been employed, for example, to determine those in a culture who are considered "wise" or who possess certain skills or abilities (Berry & Irvine, 1986).

Task Properties. In other studies the primary focus has been on the characteristics of the task or problem, rather than on attributes or competencies of the problem solver. A number of different task properties have been examined: substantive domain, importance of task, age-appropriate tasks, well- versus ill-structured tasks, and common versus uncommon tasks (Sinnott, 1989).

Tasks are most frequently characterized in terms of the substantive domain represented. Substantive domains examined include interpersonal competence and coping (Cornelius & Caspi, 1987), games and leisure activities (Clark et al., 1987; Charness, 1981, 1983; Rice et al., 1988), work-related skills (Hoyer,

1985; Salthouse, 1984; Wagner & Sternberg, 1985), and tasks of daily living (Camp et al., 1989; Willis, 1991).

There has been the assumption that everyday cognition is the study of "important" tasks or problems experienced relatively frequently by the "average" adult. In this vein, Sinnott suggests that we ask, "Is this (task) ever likely to happen to any one, and if it did, would they care?" (1989, p. 40). A cursory survey of the literature, however, suggests that among researchers studying everyday cognition, there is little consensus on what are the *most important* everyday problems and the frequency of their occurrence.

We conducted a study (Diehl & Willis, 1990) recently to examine the degree of consensus on two questions among different types of service providers for the elderly: (a) What are the most important tasks of daily living that an older adult must be able to perform in order to live independently? and (b) How frequently are various tasks of daily living encountered by older adults. Three groups of service providers (senior center directors, managers of housing for the elderly, and occupational therapists) rated 106 everyday tasks on both the importance and frequency dimensions. The 106 tasks involved printed materials representing five IADL domains (food preparation, medications, phone usage, shopping, and financial management). Three types of printed material (directions, charts/schedules, forms) were included in each domain. As shown in Fig. 3.5, there was considerable agreement among the three groups on the IADL domains considered most important for independent living. Tasks related to medications (e.g., interpreting a medicine bottle label) and financial management (e.g., comprehending a phone bill) were rated as being most important for independent living. In contrast, there was much less consensus on the dimension of frequency. Raters differed in how frequently they thought various tasks were likely to be encountered by older adults.

Context. The third approach to the study of everyday cognition has focused on the context, acknowledging that everyday problem solving does not occur

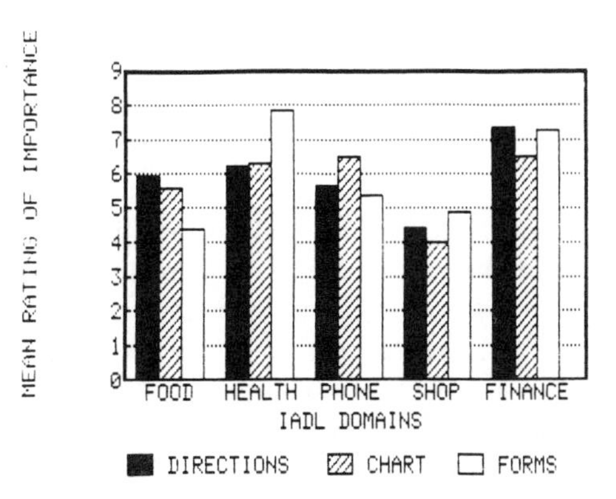

FIG. 3.5. Ratings of importance of tasks representing five domains of daily living by 3 groups of service providers (senior center directors, managers of housing for the elderly, occupational therapists).

within an environmental vacuum (Baltes, Wahl, & Schmid-Furstoss, 1990; Willis, 1991). In the broadest sense, context is defined by the cultural environment within which individuals engage in problem solving behavior. But contextual domains can be elaborated in increasingly specific ways as we consider subsets related to social class, work settings, age-structured communities, and ethnic subcultures (cf. Scribner, 1984). However, knowing the cultural expectations and the class membership of our problem solvers will not suffice in helping us understand the more specific attributes of contextual situations within which everyday problem solving is to occur. Hence, a quasi-ethnological approach is needed to discover the everyday situations in which older persons are expected to solve problems. Such an approach can yield a data language that will allow us to describe the contextual attributes of specific situations within which problem solving occurs.

A study by Scheidt and Schaie (1978) discovered that most everyday contexts within which older persons were expected to solve problems could be arranged within a four-dimensional framework. Problem solving situations were seen as occurring within a social or nonsocial context, were perceived as being common or uncommon, had supporting or depriving elements, and required either active or passive responses on the part of the older problem solver. Age difference findings revealed that with increasing age, older adults perceived themselves as more competent in situations that involved social, common, passive, and depriving features (Willis & Schaie, 1986).

Integrating the Facets. If we assume that these three approaches, focusing on personal attributes, task properties, and context, simply represent alternate facets of the everyday problem solving domain (see Fig. 3.4), it then becomes clear that study of only a single facet will not suffice. What is needed are comprehensive studies that cross elements of all three facets in a single design. Such a requirement would impose excessive demands upon a single investigator, if all possible personal attributes, task properties, and contexts were to be crossed. Nevertheless, it is not unreasonable to hope that future studies will study the interrelationship among personal attributes of problem solvers, task properties, and contextual factors.

Sampling Issues

Once the problem is defined, one then must select the sample on which to study the problem. In studies of everyday cognition, external validity is, of course, a primary concern, and thus, the issue of sampling representativeness must be considered. Questions must be asked, such as: To what populations is the study of everyday cognition to be generalized? On what critera are the samples representative? A review of the literature indicates two sampling procedures that are frequently used and that require careful consideration.

Specialized Vs. Representative Samples. The study of expertise has involved the use of samples with highly specialized skills or knowledge bases, such as master chess players, expert typists, successful business executives, and pros at the race track. Use of these unique samples is clearly appropriate when one is concerned with identifying qualitative differences between highly competent individuals and novices with respect to their knowledge base or skills. However, there are limitations in generalizing findings from these studies to more representative populations (Salthouse, 1990). Expert samples are probably not appropriate for identifying predictors or correlates of everyday cognition because of the problem of restriction of range. Almost by definition, there is a restriction in the range of talent or ability in expert samples; these subjects are likely to be functioning at very high levels not only on domain-specific tasks, but also on more general skills and abilities that are germane to their area of expertise. Many cognitive measures that have sufficient range in the general population are likely to exhibit near ceiling effects in expert samples, significantly reducing correlations and thus undermining the search for meaningful relationships with a host of variables. Typically, experts also show less variability on factors that are known to be related to cognitive problem solving, such as health and socioeconomic status.

Extreme Age Group Comparisons. Several studies of everyday cognition have sought to identify tasks on which the elderly perform as well as or better than the young (Denney & Pierce, 1989). These studies have typically involved extreme age group comparisons. Although age is presented as the primary individual difference variable, the college and older adult samples may also vary on a number of other factors, such as health, use of prescription drugs, and sensory limitations, that are known to be related to proficiency in problem solving. A major problem with extreme age group designs is that it is virtually impossible to equate representatives of young and old on all the variables required to rule out rival hypotheses for group differences (Hertzog, 1990).

Recently, researchers employing extreme age group designs have sought to "equate" the old and young on vocabulary scores. Frequently, the old are reported to have mean vocabulary scores equal to or above those of the young. Two issues are of concern in equating extreme groups on verbal ability scores. First, there is the question whether verbal ability is of particular relevance to the everyday task being studied. For example, if the everyday task involves map reading, then groups need to be equated on spatial ability rather than on verbal ability. Second, it is likely that the young and old are at different points of their developmental trajectory with respect to verbal ability, even though their mean scores may be comparable. In that verbal ability, on average, does not peak until the 50s and remains stable into old age, it is likely that one is comparing young adults who have not yet reached their peak level with older adults who have passed their peak, but have not yet exhibited appreciable decline (Schaie, 1990).

Finally, it should be noted that issues of sample representativeness and generalizability are of particular concern when studying the old-old (75–84 years) and very old (85+ years). With advancing age, increasing proportions of the elderly are institutionalized, such that approximately 20% of the very old reside in assisted living contexts. Thus, studies of everyday cognition among the community-dwelling very old will represent increasingly selective samples.

Assessing Everyday Cognition

In the final part of this chapter, the thorny issues of assessing and measuring everyday cognition are discussed. Two questions are examined: (1) What methods or procedures are useful in assessing everyday cognition, and (2) How is the adequacy of performance on everyday tasks to be judged?

In order to understand how people perform on everyday tasks, it does not suffice to remain an observer of subjects' performance in their natural habitat. The purpose of the laboratory in the behavioral sciences has always been to take a complex phenomenon and to synthesize it by purposefully introducing controls. Thus, for everyday cognition as well, complex behavior must be decomposed into its parts at successive levels of reduction. No matter how arbitrary this process may appear, it is the standard method of science. Only when the components of everyday behavior are understood is it possible to link the individual's performance on these components to the adequacy of the global behavior observed outside of the laboratory.

Methods for Assessing Everyday Cognition. Six procedures frequently employed in assessing performance on practical problems are: Naturalistic observation, interview, "think aloud" or verbal description of problem solution, paper-and-pencil measure, task simulation, and computer-interactive problems. For the most part, these are the same methods and procedures that have been used in studying traditional forms of intelligence. How effectively can the same methods and procedures be used to study everyday cognition? There is a continuum of methods that ranges in appropriateness from assessing detailed behavioral components to those offering descriptions of global phenomena. What particular method or procedure is to be used will necessarily depend on the step at which the investigator is operating. If it is the objective of a study to describe the global behavior, then there is no substitute for direct observation. To bring the behavior under control for study and to identify dimensions of the phenomenon, more structured methods are required. Once the behavior has been described, and its dimensions have been understood, the scientific objective must be to design studies that decompose the phenomenon. In the latter instance, the methods of study can no longer mirror the conditions under which the behavior occurs in the natural environment. Because the objective is no longer to describe the whole but instead to analyze its parts, it is now necessary to impose controls of a type that are not present in the naturalistic context. At this point, our more traditional procedures that allow for experimental control become the methods of choice.

Internal and External Validity. With respect to internal validity, there are the issues of convergent versus divergent validity. The question of *convergent* validity is to determine whether the dimensions of everyday cognition represent a coherent domain. If research on the construct of everyday cognition is to transcend the study of isolated tasks, then it is necessary to discover superordinate dimensions—hence the need for task taxonomies. The question of *divergent* validity involves the demonstration that the dimensions of everyday cognition can indeed be defined independently of the ability constructs measured by traditional approaches to the study of adult intelligence.

With respect to *external* validity, it is necessary to show that the factors identified through controlled studies in the laboratory can be generalized to everyday behavior in the naturalistic context. Here reference is made to our earlier definition of everyday cognition, in which three facets (personal attributes, task properties, contexts) were identified (see Fig. 3.4). It is essential to demonstrate that the laboratory-defined facets can be observed in everyday problem solving in the real world. That is, it would be essential to show that personal attributes, task properties, and context account for variability in everyday functioning in the real world, as well as in the laboratory.

Adequacy of Problem Solutions. In the final instance, what counts is whether or not an individual succeeds in solving a problem. But who will determine the adequacy of the problem solution, and how is this judgment to be made? For some everyday tasks, the problem solver's own judgment of the adequacy of problem solution is sufficient. An example is a decision regarding the purchase of household furnishings. However, even in this instance, there may be a hierarchy of external criteria that would inform us whether the criteria used by the problem solver is totally individualistic or whether it conforms to criteria more generally accepted by others.

There are other, perhaps more important everyday problems, where external criteria for judging adequacy of problem solution are required. There are at least three sources for these external criteria: Social consensus, theory, and empirical data. For example, the appropriateness of different ways of resolving interpersonal conflict is often judged by criteria defined by social consensus (Cornelius & Caspi, 1987). On the other hand, criteria for assessing the adequacy of solutions for moral dilemmas have been generated by theory (e.g., Kohlberg's stages; religious theology). Empirical criteria for assessing the adequacy of a solution, alternatively, are typically developed by experts through experimental trials on a case by case basis, such as determining the best sequence of steps in assembling a piece of equipment.

SUMMARY

This chapter has considered some of the salient taxonomic and methodological issues that confront the study of everyday cognition. Everyday problem solving has at least three defining characteristics: (a) the application of cognitive abilities and skills, (b) within naturalistic or everyday contexts, to (c) problems that are complex and multidimensional. Some of the dimensions of everyday problem solving were then examined. These include: a cognitive core of basic abilities and domain-specific knowledge bases, followed by consideration of the problem solver's individual circumstances, as well as social and affective factors. Problem solving involves the *integration* of these components. In many instances problem solving turns out to be a recursive process.

Four methodological issues were considered. First, the factors currently employed in studying the domain of everyday problem solving include: personal attributes, task properties, and contexts. Future research needs to examine the interrelationship among these factors, rather than defining everyday cognition by a single factor. Second, there are sampling issues, including the limitations of specialized samples and of extreme age group comparisons. Third, methods currently employed in the assessment of everyday problem solving were examined. Finally, the ways in which the adequacy of problem solutions are judged were discussed.

The study of everyday cognition is a challenging and fruitful endeavor. However, as this overview of methodological and taxonomic issues suggests, the complexity of the problem will require intensive and rigorous studies utilizing all of the sophistication that characterizes both the psychometric and experimental traditions in the study of cognitive development.

ACKNOWLEDGMENTS

This chapter was written with the support of research grants AG08082 and R37AG08055 from the National Institute on Aging.

REFERENCES

Baltes, M. M., & Baltes, P. B. (1986). *The psychology of control and aging*. Hillsdale, NJ: Lawrence Erlbaum Associates.

Baltes, P., Dittman-Kohli, F., & Dixon, R. (1984). New perspective on the development of intelligence in adulthood: Toward a dual-process conception and a model of selective optimization with compensation. in P. Baltes & O. Brim, Jr. (Eds.), *Life-span development and behavior* (Vol. 6, pp. 33–76). New York: Academic Press.

Baltes, M. M., Wahl, H. W., & Schmid-Furstoss, V. (1990). The daily life of elderly humans: Activity patterns, personal control, & functional health. *Journal of Gerontology; Psychological Science, 45*, 173–179.

Berg, C. (1990). What is intellectual efficacy over the life course?: Using adults' conceptions to address the question. In J. Rodin, C. Schooler, & K. W. Schaie (Eds.), *Self-directedness: Cause and effects throughout the life course* (pp. 155–182). Hillsdale, NJ: Lawrence Erlbaum Associates.

Berry, J., & Irvine, S. (1986). Bricolage: Savages do it daily. In R. Sternberg & R. Wagner (Eds.), *Practical intelligence* (pp. 236–270). New York: Cambridge University Press.

Camp, C., Doherty, K., Moody-Thomas, & Denney, N. (1989). Practical problem solving in adults: A comparison of problem types and scoring methods. In J. Sinnott (Ed.), *Everyday problem solving: Theory and applications* (pp. 211–228). New York: Praeger.

Capon, N., & Kuhn, D. (1979). Logical reasoning in the supermarket: Adult females' use of a proportional reasoning strategy in an everyday context. *Developmental Psychology, 15,* 450–452.

Capon, N., & Kuhn, D. (1982). Can consumers calculate best buys? *Journal of Consumer Research, 8,* 449–453.

Cavanaugh, J. C. (1983). Comprehension and retension of television programs by 20- and 60-year-olds. *Journal of Gerontology, 38,* 190–196.

Ceci, S., & Liker, J. (1986). Academic and nonacademic intelligence: An experimental separation. In R. Sternberg & R. Wagner (Eds.), *Practical intelligence* (pp. 236–270). New York: Cambridge University Press.

Charlesworth, W. (1976). Intelligence as adaptation: an ethological approach. In L. Resnick (Ed.), *The nature of intelligence* (pp. 147–168). Norwood, NJ: Ablex.

Charness, N. (1981). Visual short-term memory and aging in chess players. *Journal of Gerontology, 36,* 615–619.

Charness, N. (1983). Age, skill, and bridge bidding: A chronometric analysis. *Journal of Verbal Learning and Verbal Behavior, 22,* 406–416.

Charness, N. (1989). Age and expertise: Responding to Talland's challenge. In L. Poon, D. Rubin, & B. Wilson (Eds.), *Everyday cognition in adulthood and old age* (pp. 437–456). New York: Cambridge University Press.

Clark, J., Lanphear, A., & Riddick, C. (1987). The effects of videogame playing on the response selection processing of elderly adults. *Journal of Gerontology, 42,* 82–85.

Cornelius, S., & Caspi, A. (1987). Everyday problem solving in adulthood and old age. *Psychology and Aging, 2,* 144–153.

Denney, N. W., & Pierce, K. A. (1989). A developmental study of practical problem solving in adults. *Psychology and Aging, 4,* 438–442.

Denney, N. W. (1989). Everyday problem solving; Methodological issues, research findings and a model. In L. W. Poon, D. C. Rubin, & B. A. Wilson (Eds.)., *Everyday cognition in adulthood & late life* (pp. 330–551). Cambridge: Cambridge University Press.

Diehl, M., & Willis, S. L. (1990, November). *Adults' perception about the relevance of printed materials for elderly's independent living.* Paper presented at the annual meeting of the Gerontological Society of America, Boston, MA.

Fillenbaum, G. (1985). Screening the elderly: A brief instrumental activities of daily living measure. *Journal of the American Geriatrics Society, 33,* 698–706.

Friederiksen, N. (1986). Toward a broader conception of human intelligence. In R. Sternberg & R. Wagner (Eds.), *Practical intelligence* (pp. 236–270). New York: Cambridge University Press.

Garfein, A. J., Schaie, K. W., & Willis, S. L. (1988). Microcomputer proficiency in later-middle-aged and older adults: Teaching old dogs new tricks. *Social Behaviour, 3,* 131–148.

Hertzog, C. (1990, March). *Methodological issues in cognitive aging research.* Paper presented at the Third Cognitive Aging Conference, Atlanta, GA.

Hoyer, W. (1985). Aging and the development of expert cognition. In T. Shlechter & M. Toglia (Eds.), *New directions in cognitive science* (pp. 69–87). Norwood, NJ: Ablex.

Lachman, M. (1986). Locus of control in aging research: A case for multidimensional and domain-specific assessment. *Psychology and Aging, 1,* 34–40.

Lawton, M. P. (1982). Competence, environmental press, and the adaptation of older people. In M. P. Lawton, P. G. Windley, & T. O. Byerts (Eds.), *Aging and the environment: Theoretical approaches* (pp. 33–59). New York: Springer.

Lawton, M. P., & Brody, E. (1969). Assessment of older people: Self-maintaining and instrumental activities of daily living. *The Gerontologist, 9,* 179–85.

Meyer, B. J. F., Marsiske, M., & Willis, S. L. (1993). Text processing variables predict the readability of everyday documents read by older adults. *Reading Research Quarterly.*

Moss, M., & Lawton, M. P. (1982). Time budgets of older people: A window on four lifestyles. *Journal of Gerontology, 37,* 115–123.

Neisser, U. (1976). General, academic, and artificial intelligence. In L. Resnick (Ed.), *Human intelligence: Perspectives on its theory and measurement* (pp. 179–189). Norwood, NJ: Ablex.

Park, D. C. (1991). Applied cognitive aging research. In F. I. M. Craik & T. A. Salthouse (Eds.), *Handbook of aging and cognition.* Hillsdale, NJ: Lawrence ErlbaumAssociates.

Rice, G. E., Meyer, B. F., & Miller, D. (1988). Relation of everyday activities of adults to their prose recall performance. *Educational Gerontology, 14,* 147–158.

Rodin, J., Timko, C., & Harris, S. (1985). The construct of control: Biological and psychosocial correlates. In M. P. Lawton (Ed.), *Annual review of gerontology and geriatrics* (Vol. 6, pp. 3–55). New York: Springer.

Salthouse, T. (1984). Effects of age and skill in typing. *Journal of Experimental Psychology: General, 13,* 345–371.

Salthouse, T. (1987). The role of experience in cognitive aging. In K. W. Schaie (Ed.), *Annual review of gerontology and geriatrics, Vol. 7* (pp. 135–158). New York: Springer.

Salthouse, T. (1990). Cognitive competence and expertise in aging. In J. E. Birren & K. W. Schaie (Eds.), *Handbook of the psychology of aging,* (3rd ed., pp. 311–318). New York: Academic.

Schaie, K. W. (1990). Intellectual development in adulthood. In J. E. Birren & K. W. Schaie (Eds.), *Handbook of the psychology of adult development* (pp. 291–310). New York: Academic.

Scheidt, R. J., & Schaie, K. W. (1978). A taxonomy of situations for the elderly population: Generating situational criteria. *Journal of Gerontology, 33,* 848–857.

Schooler, C. (1989). Social structure effects and experimental situations: Mutual lessons of cognitive and social science. In K. W. Schaie & C. Schooler (Eds.), *Social structure and aging: Psychological processes* (pp. 131–141). Hillsdale, NJ: Lawrence Erlbaum Associates.

Scribner, S. (1984). Studying working intelligence. In B. Rogoff & J. Lave (Eds.), *Everyday cognition: Its development and social context* (pp. 9–40). Cambridge, MA: Harvard University Press.

Sinnott, J. (1989). *Everyday problem solving: Theory and applications.* New York: Praeger.

Sprafkin, R. P., & Goldstein, A. P. (1990). Using behavioral modeling to enhance professional competence. In S. L. Willis & S. Dubin (Eds.), *Maintaining professional competence* (pp. 262–277). San Francisco: Jossey-Bass.

Sternberg, R., & Berg, C. (1987). What are theories of adult intellectual development theories of? In C. Schooler & K. W. Schaie (Eds.), *Cognitive functioning and social structure over the life course* (pp. 3–23). Norwood, NJ: Ablex.

Sternberg, R., Conway, B., Keton, J., & Berstein, M. (1981). People's conceptions of intelligence. *Journal of Personality and Social Psychology, 41,* 37–55.

Wagner, R. K. (1986). The search for intraterrestrial intelligence. In R. J. Sternberg & R. K. Wagner (Eds.), *Practical intelligence* (pp. 361–378). New York: Cambridge University Press.

Wagner, R. K., & Sternberg, R. J. (1985). Practical intelligence in real-world pursuits: The role of tacit knowledge. *Journal of Personality and Social Psychology, 48,* 436–458.

Wallston, K. A., & Wallston, B. S. (1982). Who is responsible for your health? The construct of health locus of control. In G. S. Sanders & J. Suls (Eds.), *Social psychology of health and illness* (pp. 65–95). Hillsdale, NJ: Lawrence Erlbaum Associates.

Willis, S. L. (1991). Cognition and everyday competence. In K. W. Schaie (Ed.), *Annual Review of Gerontology and Geriatrics* (Vol. 11). New York: Springer.

Willis, S. L., Maier, H., & Tosti-Vasey, J. L. (1992). *Cognitive abilities, jigsaw & crossword puzzles*. Unpublished manuscript. University Park, PA: The Pennsylvania State University.

Willis, S. L., & Marsiske, M. (1991). Life-span perspective on practical intelligence. In D. E. Tupper & K. D. Cicerone (Eds.), *The neuropsychology of everyday life: Issues in development and rehabilitation* (pp. 183–198). Boston: Kluwer.

Willis, S., & Schaie, K. W. (1986). Practical intelligence in later adulthood. In R. Sternberg & R. Wagner (Eds.), *Practical intelligence* (pp. 236–270). New York: Cambridge University Press.

4 Memory in the Laboratory and Everyday Memory: The Case for Both

Eugene Winograd
Emory University

My purpose in this chapter is to argue for diversity of method in the scientific study of memory. As I have argued elsewhere (Winograd, 1988a), it is highly desirable to look for continuities in memory function across different settings. Does anyone think otherwise? I don't think so, but the controversy between advocates of an ecological or naturalistic approach to the study of memory (Bahrick, 1991; Neisser, 1978) and defenders of the primacy of the laboratory (Banaji & Crowder, 1989) has begun to suggest an unfortunate polarity. One might dismiss it and say, along with Tulving (1991) that "it is a genuine tempest in an ersatz teapot" (p. 41) that matters little, but more needs to be said in delineating the essential unity of our enterprise. My underlying theme is that polarization around questions of method is counterproductive when faced with the kinds of problems the study of memory presents.

The approach I take here has two parts. First, I address one class of reasons to venture outside the psychology laboratory to study memory; second, I attempt to relate enough converging findings to lend confidence to the proposition that, although our methods may differ according to the particular problem being studied, the same principles often underlie phenomena observed in widely different settings.

WHY LEAVE THE LABORATORY?

In the laboratory, one has control over the conditions of learning, the duration of the retention interval, and the activities occurring during it, as well as the retrieval test used. With these advantages, it is reasonable to ask, "Why leave the laboratory?" Can any possible gains offset these advantages? Although different reasons have been offered (see Bruce, 1989), I emphasize the following: There are a number of important and interesting phenomena of memory that are very hard, perhaps impossible, to study in the traditional memory laboratory. Therefore, they tend to be ignored. Because there is no shortage of worthwhile problems amenable to traditional methods, the tendency is to continue with the methods we know and trust. One result of this is that we study only tractable problems. Furthermore, the laboratory tradition inculates the belief that we are continually chipping away at the unknown. It is assumed that the more difficult problems will turn out to be special cases of phenomena we do understand. What I propose to do is to present some problems that are unlikely to be dealt with satisfactorily unless we incorporate naturalistic methods. Among these problems are involuntary remembering; the problem of "retrieval mode"; forgetting over very long intervals; autobiographical memory; remembering information learned in educational settings; mood and memory; and prospective remembering.

INVOLUNTARY REMEMBERING

A general theme of this chapter is that the memory laboratory is extremely well suited to the study of learning whereas most of the interesting problems of everyday memory concern retrieval. The case of incidental retrieval, or involuntary remembering, is a case in point. Often we remember things without trying to recollect at all; they just seem to occur to us. (Let me emphasize that I do not mean implicit memory or priming. I only include cases where there is a strong sense of pastness; it is recollection that I mean.) Ebbinghaus wrote in 1885;

> Often, even after years, mental states once present in consciousness return to it with apparent spontaneity and without any act of the will; that is, they are reproduced involuntarily. Here, also, in the majority of cases we at once recognize the returned mental state as one that has already been experienced; that is, we remember it. (p. 2)

This Proustian memory experience presumably is cued in some manner by the environment as well as by our internal state. Indeed, on some occasions, we identify the cue and say "that reminds me. . . ." More often, however, we do not know what the reminder is. We are overpowered by the memory. Others have written about this discrepancy and tried to come to terms with it. Thus, Esther Salaman (1982) engaged in an intensive analysis of her own involuntary memories and tried to classify them. Whereas Salaman relied mainly on her own involuntary memories, almost all stemming from her childhood, Spence (1988) analyzed examples from published autobiographies. Spence says that we rarely notice the "enabling context," his term for retrieval cues, when engaged in involuntary remembering. Our focus is so completely on the remembered event that we do not notice what triggers the memory. Spence believes that there is a reason for not noticing the context, for our not being able to say what instigates the act of remembering. He says, "we may recognize the memory, but be unaware of the enabling context, precisely because one of the functions of the former is to make us unaware of the latter. To gather data on both phenomena demands both special interest and special training" (p. 314).

Spence's point is that involuntary remembering ("passive memories" is his preferred term) is actually a form of motivated forgetting, if I understand him correctly. We are flooded by the memory in order to prevent us from noticing an aspect of the environment that may be threatening to us; one function of the memory is to distract us from the very aspects of the environment that give rise to the act of remembering. Strongly influenced by psychoanalytic theory, both Salaman and Spence see a major role of emotionality in involuntary remembering. It is possible that they overestimate the importance of affect and that much involuntary remembering is more mundane in nature.

Schank's (1982) analysis of reminding is replete with examples of involuntary memories taken from everyday life and constitute an interesting corpus. In Schank's case, these examples are used to support his view of memory as a dynamic, failure-driven system. In contrast to Salaman and Spence, Schank starts with concepts taken from artificial intelligence and cognitive psychology and uses everyday examples of remindings as the data base. Of interest here is that Schank's method is that of the naturalist; he collected instances of reminding from his life and the lives of those around him.

Methodologically, the problem of involuntary memories is that it is virtually impossible to study them any way but naturalistically. If you just say "Tell me some of your memories," you will obtain voluntary memories. The diary method seems all that is available to us. One observes them on the wing.

RETRIEVAL MODE: THE PROBLEM OF UNDERREMEMBERING

Related to involuntary memory is the problem Tulving (1983) addresses as "retrieval mode." Essentially, the problem is that, in a world full of retrieval cues, and with organisms whose capacity for storing memories is incredible, why is it that we don't spend our lives lost in recollection? Tulving (1983) says, "people have marvellous capacities for remembering their past, but they usually spend little time in reminiscing. Moreover, many stimuli that could potentially serve as reminders or cues, even if prominently displayed to a person, have no such effect on him" (p. 169). One might say that it would not be adaptive to be remembering constantly, that it would interfere with our going about our present business, but the conditions for remembering seem to be almost constantly present; we rarely find ourselves in a totally unfamiliar environment. Tulving's answer is to invoke the concept of "retrieval mode." Thus, "the episodic system must be in the 'retrieval mode' before a stimulus change in the environment can serve as an effective retrieval cue to stored episodic information" (p. 46). While invoking a special state of mind may not seem to advance our understanding a great deal, Tulving has focused on a real problem. It is of interest that Spence notes that passive memories may be associated with some alteration of state, as in just before falling asleep. At this point, speculation can go in many directions. But the problem of accounting for the setting factors for remembering remains.

Of course, there is no problem of inducing retrieval mode in the laboratory at all. The subject has agreed to be in a memory experiment, engages in various tasks, and then is asked to recall or recognize the material just encountered. All the setting factors for remembering are in place in the laboratory; the subject is in the retrieval mode. We cannot even see the problem unless we leave the laboratory.

FORGETTING OVER LONG INTERVALS

The naturalistic study of memory has met with considerable success in studying the retention of information over very long intervals. It has often been noted that laboratory studies of memory have had little light to shed on everyday forgetting, the problem of greatest interest to the layman. It is unusual to find retention intervals in laboratory experiments that are measured in units greater than minutes. Ebbinghaus' (1885) heroic effort is still reproduced in most textbooks. One reason for this is that Ebbinghaus used retention intervals as long as 31 days. Until recently, Ebbinghaus' function, showing forgetting at such a rapid rate that less than half of the material he had mastered was retained at even a 2-hour interval, was taken as the mirror of retention. What a leaky bucket memory appears to be. When Peterson and Peterson (1959) showed rapid forgetting was the rule even in short-term memory, with a curve mirroring Ebbinghaus' rapid decay function over 18 sec, it appeared certain that there was a universal forgetting curve.

We now know that forgetting curves are not necessarily of the leaky bucket variety and that the passage of time produces varied effects depending on what is being remembered. This discovery required leaving the laboratory. The work of Harry Bahrick is highly instructive. Bahrick, Bahrick, and Wittlinger (1975) studied the retention of the names and faces of high school classmates over a period of almost 50 years and found no evidence of forgetting at all on several of their measures over the first 35 years after graduation. How does one explain this total absence of forgetting over so long a period? Why isn't the bucket leaking? One answer might be that, after all, as everyone knows, Ebbinghaus and the Petersons studied memory for nonsense trigrams; Bahrick et al. show what can happen with socially meaningful memoria such as real people in your life. This hypothesis cannot stand up to another Bahrick (1984a) finding, that college professors are leaky buckets when it comes to remembering the names and faces of their former students. How about TV shows? Squire (1989) studied the retention of the names of TV shows that only lasted one season and never made it to reruns. Over a period of 15 years, forgetting looked neither like Ebbinghaus' logarithmic decay function nor like Bahrick et al.'s plateau of retention. Instead, Squire found a continuous but gradual forgetting curve that seems to represent an intermediate case. Although it is difficult to come up with a satisfactory theoretical account of this array of data, it is clear that forgetting has many manifestations that we would not have known about unless we looked beyond the human learning laboratory.

Bahrick has suggested that the resistance to forgetting observed for the names and faces of high school classmates reflects both the very high degrees of learning associated with people who were seen almost daily for many years as well as the favorable spacing of practice. Encounters with classmates were spaced over time. Whole summers pass without any encounters and then the process of regular interactions starts up again in accord with the academic calendar. That distributed practice facilitates long term retention is well known in the laboratory (Baddeley, 1990). Here is a case of convergence, if Bahrick is correct; principles known from laboratory study can make sense of unusual data patterns found in nature. After all, a professor's exposure to students is limited to one quarter or semester for the most part and exposure to TV shows that do not survive one season are also limited to one time period. Although these time periods are rather extensive compared to the three minutes spent studying a word list, it is reasonable to argue that the spacing is much greater in Bahrick's classmate study.

In 1984, Bahrick (1984b) published the results of his investigation of retention of Spanish that had been learned in school. He tested subjects at retention intervals as long as 50 years. The outcome was a unique forgetting curve. Forgetting was continuous for about 5 years and then stopped for about 25 years. Essentially, information that survived the 5-year reign of terror during which forgetting occurred was, as Bahrick put it, now in "permastore," immune to forgetting. Furthermore, the proportion of Spanish learned that made it into the safe haven of permastore was an increasing function of both the number of years of study and the average grade received in Spanish. (Not only does this result show the importance of degree of learning in facilitating retention, it suggests that there is some validity to the grading done by college professors.) Students of aging and memory might note that in both of Bahrick's experiments showing no forgetting over 25–35 years, there is a late drop associated with aging.

Methodologically, it is interesting to note that, in all the studies mentioned so far in this section, the investigator has had to work as a memory detective, constructing the past on the basis of the evidence at hand. Thus, Bahrick and his associates obtained the information they needed in order to construct a memory test from high school yearbooks, teacher's grade books, and university registrars. Squire obtained information about TV shows through careful analysis of public

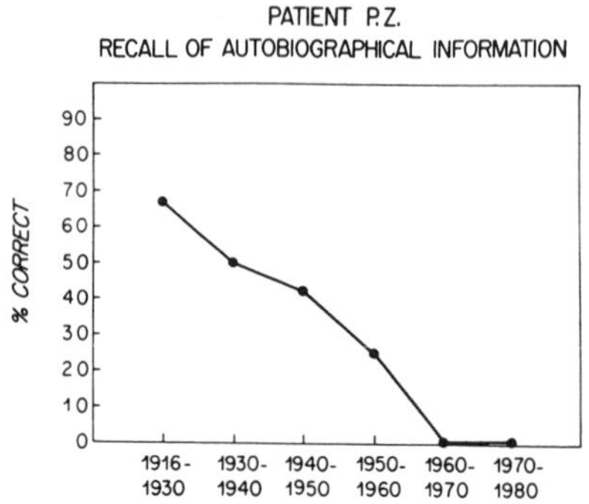

FIG. 4.1. Recall of information from his published autobiography by decade for amnesia patient P.Z. From Butters and Cermak (1986). Permission granted by Cambridge University Press.

records about TV broadcasts as well as Nielsen ratings of audience size. To reiterate a point I raised earlier, it is common for the student of naturalistic memory to enter the scene at retrieval, at the time of remembering. A frequent challenge is to come to terms with what happened at encoding. Bahrick and Squire used records of past events to guide their research. It is just this aspect of naturalistic memory research that makes laboratory oriented researchers nervous about the whole enterprise. I suggest that the choice about long term forgetting is simple: either one chooses ignorance and the safety of the laboratory or one ventures out to where the phenomenon of interest lies. The other alternative, of course, is to control learning in the laboratory and then bring back the participants over retention intervals of up to 50 years.

I cannot resist presenting one final example of a long-term forgetting function based on a single case study of memory pathology. The memory detectives here were Butters and Cermak (1986) who tested the memory of a Korsakoff patient in his 60s for information from the patient's own life. The source of the tested information, and its dates, was the patient's autobiography, written a few years before the onset of the disease. Figure 4.1 shows the remarkable retrograde amnesia manifested by P.Z. Although elderly people often complain that they remember events of many years ago more clearly than the events of yesterday (I discuss the basis for this memory illusion later), it is doubtful that any normal person, if tested, would show the inverse forgetting function shown by P.Z. He seems to remember nothing about events of the past 20 years but shows increasing access to memories the older they are. These quantitative data are in full agreement with clinical descriptions of retrograde amnesia in Korsakoff patients (see Sacks' (1987, ch. 2) account of such a patient).

AUTOBIOGRAPHICAL MEMORY RETENTION FUNCTIONS

Most studies of autobiographical memory do not have a record of what was learned. Rather, the participants are asked for personal memories in response to selected cues, usually concrete nouns, using the method developed by Galton. In addition, the participants are asked to date the memories retrieved in response to the cues. The functions obtained this way are more properly viewed as retention functions than as forgetting functions. In these studies, one can observe the temporal distribution of accessible memories. An interesting reminiscence effect associated with aging has been found in older subjects. Figure 4.2 shows the distribution of memories obtained from subjects who were approximately 70-years-old at the time of retrieval. The frequency of memories is on the vertical axis and the number of memories assigned to each decade by the subject is plotted on the horizontal axis. Rubin, Wetzler, and Nebes (1986) have plotted the results from four different studies with the top function representing the sum across the studies. Unlike the amnesic P.Z., these normal septuagenerians access more memories from the most recent decade than any other decade. However, they do show an interesting reminiscence effect; the function is nonmonotonic, showing a clear increase in accessibility for events that occurred between the ages of 10 to 30 years. The same effect has been shown for 50-year-olds by Rubin et al. (1986); that is, the 50-year-olds also show an increase in accessibility of memories associated with the second and third decades of life. Why should this be the case? I suggest that these functions tell us as much about people's lives in our culture as they do about the principles of memory. It would

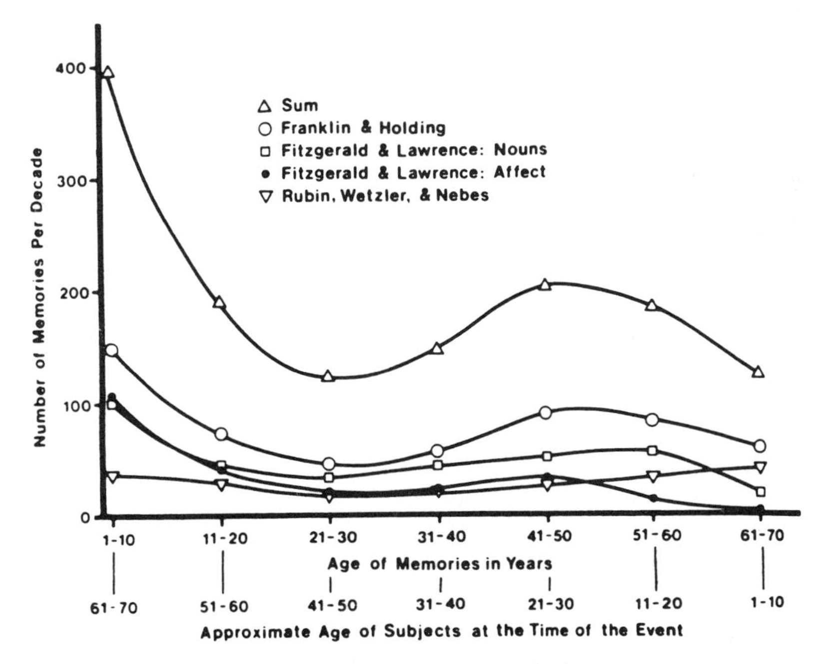

FIG. 4.2. Number of autobiographical memories by decade for 70-year-old subjects. The top curve is the sum of the four lower curves. From Rubin, Wetzler, and Nebes (1986). Permission granted by Cambridge University Press.

appear that memories associated with life transitions such as one's first job, leaving home, first romance, and the birth of children stand as landmarks as one looks back at one's life. The retention function reflects social development. Presumably, the vividness of these 50-year-old memories give rise to the illusion that one's memory for recent events is impaired with aging. It is unlikely that anything that happened last week can compete in vividness with the transitions of early adulthood. Still, the figure shows that more memories are, in fact, retrieved from the most recent decade than from any other, in spite of the illusion.

CONTINUITIES IN FORGETTING: CUE DISTINCTIVENESS

I have already emphasized that Bahrick explained the lack of forgetting in his high school classmate study in terms of overloading and distribution of practice, concepts originating in the laboratory. In this section, I show another continuity between a naturalistic phenomenon and laboratory principles. The principle is cue distinctiveness (or cue loading) and states that the efficiency of a retrieval cue declines as the number of items associated with the cue increases (Watkins & Watkins, 1975). An earlier instance of this principle was the classic list length effect; the probability of successful retrieval of an item declines the larger the set of which it is a member (Woodworth, 1938). The optimal situation obtains when the cue is unique to the information it is associated with. Classical mnemonic methods such as the Method of Loci use a one-to-one pairing of cue and to-be-remembered information. Cue distinctiveness is a powerful principle and applies to the much studied phenomena of classical interference. In studies of retroactive and proactive inhibition, there is a cue load of one for the control list and a cue load of two for the experimental condition. At retrieval, the control group is in a more favorable part of the cue loading function.

I now show how neatly the cue loading principle applies to Wagenaar's (1986, 1988) diary study of his own memory. Wagenaar recorded daily entries of events from his life over a 6-year period, taking care to record for each event information about *who* was involved, *what* the event was, and *where* the event happened. (Let me give an example using an event from my life. Suppose that a diary entry for 3 years ago says, "This evening (*when*) in Chicago (*where*) I went to hear jazz (*what*) with Darryl (*who*)"). From this corpus, Wagenaar tested his memory for events by means of different retrieval cues. (I focus here only on the tests where a single cue was presented). With respect to the potency of retrieval cues, Wagenaar (1986) found that *what* was the most potent or efficient cue, *who* and *where* were moderately efficient, and *when* was the least efficient cue by far. In his reanalysis of the data, Wagenaar (1988) demonstrates that the differences in cuing efficiency of the various aspects of the events were, in fact, attributable to cue distinctiveness.

In Fig. 4.3, I have plotted some of Wagenaar's extensive analysis of his data.

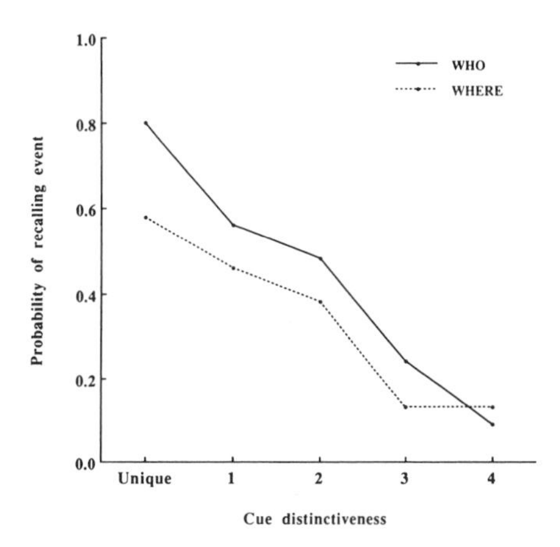

FIG. 4.3. Probability of retrieving *what,* given cue of *who* or *where,* as a function of cue distinctiveness in diary corpus. Adapted from Table 2 of Wagenaar (1988).

Shown are the probability of recalling *what* given either the cue of *who* or *where*. The functions are based on 1605 different events entered in Wagenaar's diary. It is clear that for both cues, the probability of successful recall of the event or activity in question declines precipitously as the distinctiveness of the cue decreases (or as the cue loading increases). In terms of my example, the more events in my memory involving Darryl, the less likely I would be to retrieve a specific target memory with his name as a cue. Functions resembling this, as noted before, are to be found in the laboratory (see Earhard, 1967). Wagenaar is also able to show that the loss of efficiency for *who,* but not *where,* interacts with the retention period. This intriguing finding begs replication, and, if replicated, explanation. Furthermore, the loss of cue potency with time for *who* is less steep when it is distinctive than when it is overloaded.

To repeat, Wagenaar was able to show that a substantial portion of his findings with respect to the potency of retrieval cues makes sense in terms of what we already know about retrieval dynamics from controlled laboratory study. What lessons should be drawn from this? One lesson is that the cue distinctiveness principle is strengthened by this demonstration of its validity in a naturalistic setting. Its external validity is clear; it can be generalized. Another lesson is that order can be found outside the laboratory and with a single subject. In addition, some of Wagenaar's findings, like Bahrick's finding of stable retention over very long periods, are not simply extensions of what we know, but discoveries. For another instructive example of cross-fertilization between laboratory and naturalistic studies, I now turn to the study of mood and memory.

MOOD AND MEMORY

Mood congruent recall refers to the observation that retrieval is facilitated for memories whose emotional valence is consistent with one's current mood. Theoretically, it is, if anything, overdetermined. Tulving's (1983) encoding specificity principle and Bower's (1981) network model of emotion incorporate such effects and mood congruent recall is seen as extremely important clinically. Current cognitive views of depression (Beck, 1976) are essentially "vicious cycle" models in which a depressed state leads to unhappy memories which in turn make one more depressed. Still, it has been pointed out that such models are too strong: The slightest tilt in one's current mood should lead in short order to euphoria or suicidal depression. To counteract such extreme mood swings, it has been suggested that people regulate their moods by engaging in mood incongruent recall (Blaney, 1986), but the evidence for mood incongruent recall has been lacking in laboratory research. However, in a recent series of experiments, Parrott and Sabini (1990) have shown, in an instructive way, how mood incongruent recall can be demonstrated using both naturalistic and laboratory studies. In their first two studies, Parrott and Sabini allowed naturally occurring events to determine mood. In one study, mood was influenced by the grades on a returned exam; in the second naturalistic study, mood was influenced by the prevailing weather, sunny or rainy. The autobiographical memory task was to remember three events from high school. Mood incongruent recall was obtained in both cases for the first memory retrieved; thus, those who did well on the exam recalled less happy experiences than did those who did poorly, and happier memories were recalled on rainy days.

Parrott and Sabini's next three studies were done in the laboratory and attempted to clear up why they found mood incongruent recall when so many other investigators have found mood congruence. The first hypothesis they considered was that the crucial factor was the setting, field versus laboratory. Their choice of mood induction technique in the laboratory was to play happy or sad music, a manipulation recently shown by Eich and Metcalfe (1989) to be effective in producing mood dependent memory (where mood at both encoding and retrieval are manipulated and recall is best when moods match). Utilizing musical mood induction, Parrott and Sabini were able to find mood incongruent recall again, but only when subjects were unaware that their mood was being manipulated. They concluded that mood incongruent recall can only be found when subjects do not perceive the relevance of their mood to the memory task. Typically, experimenters have to go to a lot of trouble to induce mood changes in their subjects in the laboratory and, presumably, subjects perceive that the experiment is about mood and tend to go along with the manipulation by retrieving matching memories. That is, there is a correlation between artificial mood induction and subjects being aware of the purpose of the experiment. Because artificial mood induction usually occurs in laboratories, it took leaving the laboratory, at least for a while, for Parrott and Sabini to find mood incongruent recall. Then, armed with this finding, they were able to return to the laboratory and isolate the difference between the field and laboratory settings that accounted for the different findings. The moral of the story is not that laboratory research is misleading; all research is potentially misleading. If Parrott and Sabini are correct, we can investigate the relationship between mood and memory in either setting, preferably both, armed with the knowledge that the subject's perceptions of the experiment are very important.

PROSPECTIVE REMEMBERING

Prospective memory is often defined as remembering to remember. It may also be defined as remembering to get things done at the appropriate time. Examples are remembering to call your mother on her birthday, to pick up wine and cheese on your way home, or to tell your spouse that her friend called while she was out. The example of the telephone message is instructive. Remembering the contents of the telephone message requires retrospective memory, as in a free recall experiment. It requires one to remember the contents of an episode. Remembering to tell your spouse that *anyone* called is an act of prospective memory. The socially important aspect of the interaction is informing her that she received the call; failure to do so is embarrassing or, as Munsat (1966) points out, if habitual is seen as a moral defect in one's character. Presumably, the failure is seen as moral, rather than cognitive, because of the social nature of the remembering. Prospective remembering is manifested in action; one mails a letter or appears for an appointment. Consistent failures in retrospective remembering can be regarded as an eccentricity or even a beloved personal habit. Marital partners frequently correct each other about shared memories. Was it in 1971 or 1972 that we went to California? But failing to pick up your spouse at the airport at the designated time is not regarded with amusement.

For researchers, the chief problem in studying prospective remembering is that it is often hard to identify or manipulate a retrieval cue. One has to remember to remember, after all. Therefore, asking the subject "what should you do now?" undercuts the most important aspect of prospective remembering. In this way, prospective remembering is much like involuntary remembering; it is difficult to specify a retrieval cue in advance. Furthermore, like involuntary memory, prospective memory is embedded in everyday life. Indeed, I have argued elsewhere (Winograd, 1988b) that prospective remembering does not lend itself to treatment as an isolable act of cognition but, rather, needs to be understood as bound up with attention, motivation, compliance, and goal orientation as part of ongoing action (see Meacham, 1988, for a similar perspective).

Developmentally, prospective remembering might have priority in early childhood because it is a means to an end. Reminiscence is more appropriate later in development; schooling places great demands upon retrospective memory.

Recently, researchers have begun to tackle prospective memory. The challenge has been to devise tasks that capture the phenomenon. One approach has been to assign tasks to subjects to be completed at a later time outside the laboratory, such as mailing a postcard to the investigator or telephoning at an appointed time (Moscovitch & Minde, cited in Moscovitch, 1982; West, 1988). Interestingly, the elderly have performed as well as or better than the young on these tasks, perhaps because they rely more on external aids such as calendars and appointment books than do the young. Laboratory studies have included such tasks as asking children to remember to take cupcakes out of the oven (Ceci & Bronfenbrenner, 1985), asking adults to remember to ask the experimenter for a red pencil when asked to draw geometric forms later on (Dobbs & Rule, 1987), and asking older adults to press a special response key if a particular target word appeared amongst words being shown as part of a short-term retrospective memory study (Einstein & McDaniel, 1990). Note that the laboratory tasks bear a strong resemblance to studies of vigilance and attention, that is, "when such and such happens, do such and such." Indeed, temporal coordination of action and events is part of prospective remembering. Often, it is too late to fulfill one's obligation. One striking finding of this small body of research is that there is no correlation between performance on retrospective and prospective memory tasks (see Einstein & McDaniel, 1990; Meacham & Leiman, 1982); in fact, Wilkins & Baddeley (1978) report an inverse correlation between performance in free recall and a prospective memory task designed to be an analogue of taking medicine. In short, people who are good at standard laboratory tasks of retrospective remembering such as free recall or recognition memory are no better at remembering to push the button at the right time than people who are not so good at free recall.

The open question is whether the recent laboratory studies have external validity; are they in fact assessing prospective remembering? Although one cannot answer this question adequately at present, I would maintain that the enterprise is worthwhile. For one thing, studying the process forces one to think about it analytically. Thus, Einstein & McDaniel (1989), in considering the absence of an age decrement in their prospective memory tasks, point out that a distinction between event-based and time-based tasks may be useful. Event-based tasks, like the one they used, require that an action be performed when a specific cue is presented, that is, "when your wife comes home, tell her that her secretary called," whereas time-based tasks, such as taking the cake out of the oven, are tied to the passage of time.

How important is the fact that, in the laboratory studies of Einstein and McDaniel (1990) and Dobbs and Rule (1987), the prospective task is embedded in a highly structured laboratory situation controlled by the experimenter? It is the experimenter who, for what must be perceived as arbitrary reasons to the subject, says "first, we will do this, and then, when I ask you to draw a circle, your job is to ask me for a red pencil." The subject is in the retrieval mode; he or she is alert and vigilant. Prospective remembering tasks done outside the laboratory, however, involve long retention intervals and are embedded in the events of daily life. Typically, one is not in the "retrieval mode."

The study of prospective memory is at the intersection of naturalistic and laboratory approaches to memory. The possibility for progress on this problem depends on contributions from both methodologies as well as the realization that, aside from requiring various methods, the problem is not simply one of memory. It will have to be conceptualized in broader terms (see Meacham, 1988; Winograd, 1988b) because it involves action in the world more directly than does retrospective remembering, which, after all, can be done quite comfortably alone in a quiet room.

CONCLUSIONS

Underlying this chapter is the view that the context of discovery is not important. The act of remembering is so complexly determined that a variety of approaches will be required if we are to understand it. From this perspective, the recent debate pitting laboratory and naturalistic research against each other is unproductive. Rejection of research on memory because it was done in the laboratory is as silly as rejection of work because it was not done in the laboratory. The approaches are complementary. Furthermore, what I have referred to here as "naturalistic" covers a variety of methodologies that may differ from each other in important ways. Certainly a diary study of one person's memory differs in many ways from a study of retention of material learned in school in which the investigator uses educational archives for evidence of degree of learning, and both of these methods differ profoundly from Freud's (1982) brilliant analysis of an early childhood memory reported in Goethe's autobiography.

If there is an underlying issue in the debate other than wounded feelings and worship of method, it is the problem of control. It is undeniable that the laboratory affords control over encoding, retention interval, materials, and other important factors. Furthermore, the last decade has shown that imaginative researchers can investigate fundamental, and new, questions without leaving the laboratory. Thus, Marcia Johnson's research on reality monitoring approaches the question of how we can discriminate an actual past event from an imagined event, for example, did I actually mail the letter or only intend to do it (Johnson & Raye, 1981)? Larry Jacoby and his colleagues have carried out a series of studies on sources of the subjective experience of familiarity (Jacoby, Kelley, & Dywan, 1989). These are but two examples of how important questions about memory continue to be approached within the laboratory. However, the emphasis here has been on important questions that are less amenable to laboratory study and therefore require other methods.

Throughout, I have emphasized continuities and convergence, showing how findings outside the laboratory are compatible with what we have learned from laboratory research. Such principles as distribution of practice, overlearning, and cue distinctiveness have been invoked in discussing naturalistic memory findings. At this point, it is appropriate to recognize that naturalistic research is more than a handmaiden to the laboratory. Demonstrating that the principles discovered in the laboratory have generality to complex settings is an important contribution, but naturalistic memory research has other functions as well. Its major function is discovering new phenomena not anticipated on the basis of what we have learned from the laboratory. Some new phenomena that have been discussed here are the stable forgetting functions over very long periods documented by Bahrick, reminiscence in autobiographical memory, and mood incongruent recall. Another finding of naturalistic memory research not discussed here is Ross' (1989) documentation that recall of personal facts such as the severity of past pain is jointly determined by one's present state along with one's implicit theory of consistency over time. Yet another purpose served by naturalistic memory research that has been emphasized by Baddeley (1988), Bruce (1985, 1989), and Neisser (1988) is to force us to inquire into the functions that memory serves. To raise this question at all is to call for observation in natural settings. In the final analysis, the criterion for judging research, surely, cannot be how it was done but rather, as I have argued elsewhere (Winograd, 1988a), whether it enhances our understanding of memory. The phenomena of remembering are so diverse that the ways of studying it must inevitably reflect that diversity.

REFERENCES

Baddeley, A. D. (1988). But what the hell is it for? In M. M. Gruneberg, P. E. Morris, & R. N. Sykes (Eds.), *Practical aspects of memory: Current research and issues, Vol. 1* (pp. 3–18). Chichester: Wiley.

Baddeley, A. D. (1990). *Human memory.* Boston: Allyn and Bacon.

Bahrick, H. P. (1984a). Memory for people. In J. E. Harris & P. E. Morris (Eds.), *Everyday memory, actions and absentmindedness* (pp. 19–34). New York: Academic Press.

Bahrick, H. P. (1984b). Semantic memory in permastore: 50 years of memory for Spanish learned in school. *Journal of Experimental Psychology: General, 113,* 1–29.

Bahrick, H. P. (1991). A speedy recovery from bankruptcy for ecological memory research. *American Psychologist, 46,* 76–77.

Bahrick, H. P., Bahrick, P. O., & Wittlinger, R. P. (1975). Fifty years of memories for names and faces: A cross-sectional approach. *Journal of Experimental Psychology: General, 104,* 54–75.

Banaji, M. R., & Crowder, R. G. (1989). The bankruptcy of everyday memory. *American Psychologist, 44,* 1185–1193.

Beck, A. T. (1976). *Cognitive therapy and the emotional disorders.* New York: Meridian.

Blaney, P. H. (1986). Affect and memory: A review. *Psychological Bulletin, 99,* 229–246.

Bower, G. H. (1981). Mood and memory. *American Psychologist, 36,* 129–148.

Bruce, D. (1985). The how and why of ecological memory. *Journal of Experimental Psychology: General, 114,* 78–90.

Bruce, D. (1989). Functional explanations of memory. In L. W. Poon, D. C. Rubin, & B. E. Wilson (Eds.), *Everyday cognition in adulthood and later life* (pp. 44–58). Cambridge, UK: Cambridge University Press.

Butters, N., & Cermak, L. S. (1986). A case study of the forgetting of autobiographical knowledge: Implications for the study of retrograde amnesia. In D. C. Rubin (Ed.), *Autobiographical memory* (pp. 253–272). Cambridge, UK: Cambridge University Press.

Ceci, S. J., & Bronfenbrenner, U. (1985). Don't forget to take the cupcakes out of the oven: Strategic time-monitoring, prospective memory, and context. *Child Development, 56,* 175–190.

Dobbs, A. R., & Rule, B. G. (1987). Prospective memory and self-reports of memory abilities in older adults. *Canadian Journal of Psychology, 41,* 209–222.

Earhard, M. (1967). Cued recall and free recall as a function of the number of items per cue. *Journal of Verbal Learning and Verbal Behavior, 6,* 257–263.

Ebbinghaus, H. (1964). *Memory: A contribution to experimental psychology.* New York: Dover. (Original work published 1885).

Eich, E., & Metcalfe, J. (1989). Mood dependent memory for internal versus external events. *Journal of Experimental Psychology: Learning, Memory, and Cognition, 15,* 443–455.

Einstein, G. O., & McDaniel, M. A. (1990). Normal aging and prospective memory. *Journal of Experimental Psychology: Learning, Memory, and Cognition, 16,* 717–726.

Freud, S. (1982). An early memory from Goethe's autobiography. In U. Neisser (Ed.), *Memory observed* (pp. 64–72). San Francisco: Freeman.

Jacoby, L. L., Kelley, C. M., & Dywan, J. (1989). Memory attribution. In H. L. Roediger, III, & F. I. M. Craik (Eds.), *Varieties of memory and consciousness: Essays in honour of Endel Tulving* (pp. 391–422). Hillsdale, NJ: Lawrence Erlbaum Associates.

Johnson, M. K., & Raye, C. L. (1981). Reality monitoring. *Psychological Review, 88,* 67–85.

Meacham, J. A. (1988). Interpersonal relations and prospective remembering. In M. M. Gruneberg, P. E. Morris, & R. N. Sykes, (Eds.), *Practical aspects of memory: Current research and issues, Vol. 1* (pp. 354–359). Chichester: Wiley.

Meacham, J. A., & Leiman, (1982). Remembering to perform future actions. In U. Neisser (Ed.), *Memory observed* (pp. 327–336). San Francisco: W. H. Freeman.

Moscovitch, M. (1982). A neuropsychological approach to memory and perception in normal and pathological aging. In F. I. M. Craik & S. Trehub (Eds.), *Aging and cognitive processes* (pp. 55–78). New York: Plenum Press.

Munsat, S. (1966). *The concept of memory.* New York: Random House.

Neisser, U. (1978). Memory: What are the important questions? In M. M. Gruneberg, P. E. Morris, & R. N. Sykes (Eds.), *Practical applications of memory* (pp. 3–24). London: Academic Press.

Neisser, U. (1988). Time present and time past. In M. M. Gruneberg, P. E. Morris, & R. N. Sykes (Eds.), *Practical aspects of memory: Current research and issues, Vol. 2* (pp. 545–560). Chichester: Wiley.

Parrott, W. G., & Sabini, J. (1990). Mood and memory under natural conditions: Evidence for mood incongruent recall. *Journal of Personality and Social Psychology, 59,* 321–336.

Peterson, L. R., & Peterson, M. J. (1959). Short-term retention of individual verbal items. *Journal of Experimental Psychology, 58,* 193–198.

Ross, M. (1989). Relation of implicit theories to the construction of personal histories. *Psychological Review, 96,* 341–357.

Rubin, D. C., Wetzler, S. E., & Nebes, R. D. (1986). Autobiographical memory across the lifespan. In D. C. Rubin (Ed.), *Autobiographical memory* (pp. 202–224). Cambridge, UK: Cambridge University Press.

Sacks, O. (1987). *The man who mistook his wife for a hat.* New York: Harper & Row.

Salaman, E. (1982). A collection of moments. In U. Neisser (Ed.), *Memory observed* (pp. 49–63). San Francisco: W. H. Freeman.

Schank, R. C. (1982). *Dynamic memory.* Cambridge, UK: Cambridge University Press.

Spence, D. P. (1988). Passive remembering. In U. Neisser & E. Winograd (Eds.), *Remembering reconsidered: Ecological and traditional approaches to the study of memory* (pp. 311–325). Cambridge, UK: Cambridge University Press.

Squire, L. P. (1989). On the course of forgetting in very long-term memory. *Journal of Experimental Psychology: Learning, memory, and Cognition, 15,* 241–245.

Tulving, E. (1983). *Elements of episodic memory.* New York: Oxford University Press.

Tulving, E. (1991). Memory research is not a zero-sum game. *American Psychologist, 46,* 41–42.

Wagenaar, W. A. (1986). My memory: A study of autobiographical memory over six years. *Cognitive Psychology, 18,* 225–252.

Wagenaar, W. A. (1988). People and places in my memory: A study of cue specificity and retrieval from autobiographical memory. In M. M. Gruneberg, P. E. Morris, & R. N. Sykes (Eds.), *Practical aspects of memory: Current research and issues, Vol. 1* (pp. 228–333). Chichester: Wiley.

Watkins, O. C., & Watkins, M. J. (1975). Buildup of proactive inhibition as a cue-overload effect. *Journal of Experimental Psychology: Human Learning and Memory, 104,* 442–452.

West, R. (1988). Prospective memory and aging. In M. M. Gruneberg, P. E. Morris, & R. N. Sykes (Eds.), *Practical aspects of memory: Current research and issues, Vol. 2* (pp. 119–125). Chichester: Wiley.

Wilkins, A. J., & Baddeley, A. D. (1978). Remembering to recall in everyday life: an approach to absentmindedness. In M. M. Gruneberg, P. E. Morris, & R. N. Sykes (Eds.), *Practical aspects of memory* (pp. 27–34). London: Academic Press.

Winograd, E. (1988a). Continuities between ecological and laboratory approaches to memory. In U. Neisser & E. Winograd (Eds.), *Remembering reconsidered: Ecological and traditional approaches to the study of memory* (pp. 11–20). New York: Cambridge University Press.

Winograd, E. (1988b). Some observations on prospective remembering. In M. M. Gruneberg, P. E. Morris, & R. N. Sykes (Ed.), *Practical aspects of memory: Current research and issues, Vol. 1* (pp. 348–353). Chichester: Wiley.

Woodworth, R. S. (1938). *Experimental Psychology,* New York: Holt.

III THEORY AND METATHEORY FOR LIFE-SPAN EVERYDAY COGNITION

5 Yes, It's Worth the Trouble! Unique Contributions From Everyday Cognitive Studies

Jan D. Sinnott
Towson State University

The Twelfth Conference on Lifespan Development was assembled to discuss mechanisms of everyday cognition. It has taken effort to convene this group. But even greater effort is needed for us to do everyday cognitive research. It's harder to study everyday thought! The variables in everyday studies prove harder to conceptualize and operationalize reliably. It is hard to decide what is "naturalistic" or "everyday," even in a certain context. Everyday studies (because of their newness) lack the track record laboratory measures have established in psychology; investigators cannot create one quickly because by their very nature tasks are valid only in specific contexts, not all contexts. Everyday studies often meet opposition because their creators do not use the generally accepted mechanistic model of human behavior, and, most frightening of all, implicitly challenge some of our most sacred world views as 20th century cognitive scientists. The list of difficulties could go on indefinitely. Yet few of us who do such research can easily say why it is that we bother to make this effort. We seem sure that we should do it—but why?

In Castaneda's (1981) book "A Separate Reality" the sorcerer don Juan reminds Carlos (that altered-state exploring anthropologist) that, after all, anything is one of a million paths . . . a path is only a path, and there is no affront to oneself or others in dropping it. . . . All paths are the same; they lead nowhere. . . . Look at every path closely and deliberately. . . . Then ask yourself and yourself alone one question. This question is one that only a very old man (sic) asks: Does this path have a heart? If it does, the path is good. If it doesn't, it is of no use. Both paths lead nowhere . . . (but) one makes for a joyful journey (while) . . . the other will make you curse your life.

I have become convinced our path does have a heart, and so I intend to continue a joyful journey on it. The heart of these studies is: to ask better, richer, juicier, more meaningful questions. That's why it's worth the trouble. The heart of the question and the uniqueness of everyday studies is the often neglected variable of *choice*: the investigator's choice of parameters to study; adaptive choice for the person or "subject"; choice within a theory among a larger number of questions and sub-theories; our discovery within everyday cognition of the knower's very real choice about cognitive "truth"; and knowing that the knower's and the experimenter's choices about "reality" and consequent creation of reality do matter immensely to reality itself. Choice creates cognitive filters. Everyday studies finally are about the knower as a creator of reality, working jointly with other creators, and about the extension of the discipline of psychology full circle, in a complex way, so that it can gain objectivity about its roots in philosophy, phenomenology, and religion. They are about a choice to change the conceptual and philosophical framework of our discipline to one that states that there are no entirely externally absolute frameworks at all, a shift such as that which hit physics in our century.

The goal of this chapter is to outline more fully what some of the better questions are, to show how these better questions occur in other new and developing branches of science, and to show how the processes elucidated by those questions link up with the tasks of lifespan development during midlife. I also hope to illustrate the search for such processes in my own research. We as a field will then be in a better position to discuss the potentially unique contributions of everyday cognitive studies.

BETTER QUESTIONS

What "better questions" does the study of everyday cognition allow us to ask? What agenda can everyday cognition give us?

One immensely different set of questions relates to *focus on process,* on the dialogue rather than the monolog, on the melody rather than the chords. Like don Juan entreating Carlos, everyday research almost begs us to focus on the path rather than on the destination. This is a benefit because an essential interest of psychology, after all, is in change from state to state, rather than on the states alone, although traditional studies usually focus on state. Some of the larger process questions are: How do people change or develop? How should we train agents of change to do that effectively? How can we get people to change their knowledge state, that is, to learn? What form does that going from "here" to "there" take? What kinds of questions are we asking in the experimental situation? How are those questions related to adaptive real-life behavior? In the everyday cognitive world, nothing lasts for long, and process is our most important product. Process may or may not be progress; but it is definitely different from state. The study of process can yield *both* basic and applied laws of psychology.

A second benefit to doing everyday cognitive studies is to set up a situation where we must *make a conscious choice about the level of research question to address,* forcing us to acknowledge and work with the idea that any one paradigm or view of truth is not the only one. The fact that everyday cognitive research is so very complicated pulls our attention to narrowing that complexity to a meaningful but tolerable level, to choosing a "truth." Few can help us decide how to narrow it; we must choose alone. Everyday cognitive studies say to us, echoing Thomas Wolfe (1942), that if we try to do them we can't "go" home again to a simple view of truth, because home is either nowhere or everywhere around us. When we are forced to choose we become better scientists because we see more clearly our own power in creating paradigms and reality. Acknowledging choice frees us creatively as scientists.

A third benefit is to question the *mechanisms of application* of a general principle to a specific, concrete, life experience. Notice that this approach differs in two ways from strictly laboratory research: (a) it broadens the search for basic cognitive processes by including other overlooked aspects of the cognitive event (enlarges the problem space); and, (b) it gives mechanisms for explaining the annoying, probabilistic, less-than-100% nature of occurrence of most basic cognitive processes. This is higher order basic research, whatever critics might argue.

A fourth set of questions *addresses functionality and compensation mechanisms.* Researchers have uncovered many processes that are a part of human thought, but have not yet connected many of those processes to how well or how poorly humans get along in the world. The conceptual components of practical intelligence are just beginning to be described (e.g., Sternberg, Wagner, and Okagaki, this volume). Some of one's cognitive processes inevitably fail at one time or another across the lifespan, but life goes on for the adaptive person. We have yet to do many studies on how individuals use their skills to make up for those failures and go on. Everyday cognitive studies are ideal for this task, on a basic and applied level.

Finally, since development over the lifespan involves an individual's applying general adaptive intelligence to his or her own unique life-over-time, a fifth benefit from studying mechanisms of everyday cognition can be to *create a better set of theories about tasks and growth in midlife and old age* by answering questions about those elements. A number of midlife development theories exist, but they do not make much use of cognitive variables, especially complex ones. After all, adaptive cognition was not expected to develop beyond adolescence (before we began to study *post*formal operations), and compensations and declines weren't thought to be that exciting. But ironically many sciences outside psychology are beginning to note the impact of cognition, of the observer, on reality over time. The knowing observer discussed in new physics, for example, is shaping his or her quantum reality. Coming back to psychology, one's cognitive filters for reality could determine, for example, the course of one's interpersonal relations over a lifespan, since those relations are based on those filters. How might theories of adult lifespan development include longterm everyday-reality-creating cognitive concepts?

Notice that the underlying theme in these five special contributions or sets of questions of everyday cognitive studies is that of ongoing *process* and *choice* in a *complex context* with a *history*. This theme involves being analytical about life as a whole, nonfragmented cognitive process. We see in the next section that current practitioners of other sciences are asking some similar thematic questions and are beginning to appreciate similar special contributions, and that developmental psychologists are beginning to think in this way. Are we all beginning to be aware of the same important basic laws?

EVERYDAY COGNITION SHARES BETTER QUESTIONS WITH OTHER MODERN SCIENCES

This section is an attempt to examine very briefly four "new" sciences as to ways they address the themes that appear in everyday cognitive research. A preliminary connection between new sciences and everyday cognition concepts is in Fig. 5.1, with additional links to other cognitive concepts and midlife development issues. We focus on: new physics and quantum theory, especially concerning the nature of reality; the new biology, especially theories of cooperative evolution and brain development; new mathematics, especially chaos theory; and new cognitive sciences, especially theories of self-referential Piagetian postformal thought. One key idea of each is discussed as an example of the potential richness of the approach and its links to everyday cognition themes. We argue that one main theme that links the new sciences is this: social, physical, or personal reality is partly constructed by the knower's choices as it is known, through principles of emergent structures. Those who can embrace this dialogic quality of reality are empowered to live fully and adaptively in this time of rapid change. Understanding these processes is basic science, even though the processes are everyday ones.

New Physics Models

New physics ideas (e.g., Wolf, 1981; Sinnott, 1981, 1984) have made a tremendous impact on science, technology, and philosophy. New physics concepts are difficult and describe the less familiar "big picture" reality. (Newtonian physics concepts, though limited, describe small scale local reality.) In times of change, or when one is trying to bring about change, the breadth of the big picture is

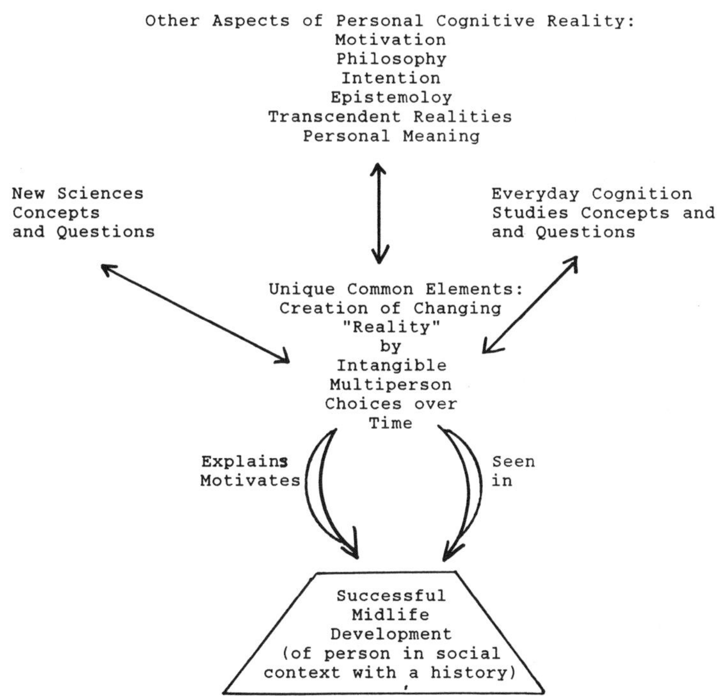

FIG. 5.1. A larger view of cognition: Relations among new science concepts, everyday cognition, broader cognitive concepts, and mid-life development issues. From Sinnott (1984). Reproduced with permission of Greenwood Publishing Group, Westport, CT.

needed in research questions since small scale descriptions have proved inadequate. But new physics ideas seem somewhat alien to us because they don't concur with our Western shared cultural myths about reality (Campbell, 1988). Until we are motivated by desperation, curiosity, or cognitive shifts to explore multiple views of reality, we are wise to avoid the challenge of new physics ideas.

In earlier and ongoing work, I describe new physics ideas related to developmental psychology, change, teaching, creativity and cognition (e.g., Sinnott 1981, 1984, 1986c, 1987d, 1989c, 1989d, 1989g, 1989h, in press-a,b,c,d; in preparation). A table summarizing shifts in world views from Newtonian to new physics ideas is reprinted here (Table 5.1). Notice that the shifts have huge implications for psychological reality. The nature of existence (in psychological terms, identity), time (lifespan development), causality (personal action, power, ability) all shift. As an example, Table 5.2 (Sinnott, 1984) describes these two world views (new and old physics) in terms of one behavioral area, interpersonal relations.

TABLE 5.1
Old Physics/New Physics Concepts

Old	*New*
Space in Euclidean.	Space is non-Euclidean, except to small regions.
Time and space are absolute.	Time and space are relative and better conceptualized as the space/time interval.
Space in uniform in nature.	Space is composed of lesser and greater resistances.
Events are located topologically on a flat surface.	Events are located topologically on the surface of a sphere.
Undistrubed movement is on a straight line.	Undisturbed movement is on a geodesic, i.e., by the laziest route.
Events are contiguous.	Events are discontinuous.
No region of events exists which cannot be known.	Unknowable regions of events exist.
Observed events are stable.	Observed events are in motion, which must be taken int account in the observation.
Formation of scientific postulates proceeds from everyday activity through generalizations based on common sense, to abstractions.	Formation of scientific postulates also includes a stage characterized by resolution of contradictions inherent in the abstractions.
Causality is deterministic.	Causality is probabilistic, except in limited space/time cases.
Cause is antecedent to and contiguous with effect.	Cause is antecedent and contiguous to event only in limiting cases. When events are grouped about a center, that center constitutes a cause.
Egocentrism is replaced by decentration during development of scientific methods.	Egocentrism and decentration are followed by taking the ego into account in all calculations.
Concepts in natural laws conform to verbal conventions.	Concepts in natural laws may appear contradictory in terms of verbal conventions.
Universe is uniform.	Universe is nonuniform–either because it is continually expanding or because it is continually being created and negated.

One conclusion drawn from the new physics is that sometimes multiple contradictory views of truth are all "true" simultaneously, although they appear contradictory at first, and that reality is therefore the view of truth to which we make a "passionate commitment" (Perry, 1975; Polanyi, 1971). Ideally, this commitment is done in awareness, with consciousness. We know that no one view of reality is, in Bronowski's (1976) words, the "God's eye view"; they are all limited by one's chosen vantage point, history, or measuring tool. This argues that anything is known only within a region of tolerance, of error, but not absolutely. If we *share* a vantage point—and only then—we share a reality. To create another personal or social reality (within limits), we need to change vantage points as individuals or groups. For example, in the physical world, from a small scale local space vantage point, parallel lines never converge; change vantage points to universal space and parallel lines always converge. Vantage point over time, or trajectory already established, also determines the event.

The impact of a simple and profound idea like that of new physics in the world of cognition can be monumental, especially for cognition between two or more social learners, and everyday cognition studies capture these questions. We could be speaking of classroom behavior, therapists confronting a changing person, development programs in third-world countries, midlife development of a couple, or interpersonal relations studies. From a new physics viewpoint, and in everyday cognition, different incompatible truths are not necessarily to be narrowed to one correct truth; they may each have their own correct logic. From a new physics viewpoint learning the truths of others can teach us greater flexibility and give us more tools for working with reality, and interpersonal cognition demands we learn others' truths. From a new physics viewpoint the line between two knowers may be a vague one since several truths are valid, and dialogue is necessary among knowers. These are all ideas of many who do everyday cognitive studies. Awareness of the new physics idea and everyday cognition idea that "truth" is partly a choice of vantage points around which we build our reality lets the classroom teacher, or Peace Corps volunteer, or negotiator, or student of aging, or therapist begin to recognize in others the power to

TABLE 5.2
Applications: Interpersonal Relations

Relativistic
Our relations are logical within a set of "givens" that we choose to utilize.
They are based on both our past relations to each other and our relations to other significant persons.
Refating means knowing "where you're coming from" and interacting on that level.
Relating is never knowing "YOU" completely, because in knowing you I am necessarily "creating" you.
Relations are always "in process"; cannot be described as stable until they end.
Formalistic
There is only one way to structure our relationship to reflect reality.
Our relationship exists "out there" in reality.
Our relationship involves only us, now.
The relationship has just one "reality" – no need to match levels to understand
We can know the essence of each other.
Role is more important than process.

From Sinnott (1984).

construct and experience—and be responsible for—their intellectual lives. Accepting shifts in reality, bridging, and dialoguing between two "truths" is likely to lead to more permanent, useful, adaptive mutual learning, which is not sabotaged by a rigid world view during times of change.

New Biology Models

Proponents of the new biology (e.g., Augros & Stanciu, 1987; Maturana & Varela, 1988; McLean, 1988) go to the original data on which evolutionary theory was formed and add the new experimental data of modern medicine and biology to attempt to answer new and difficult biological questions. In doing so they come to conclusions that stand evolutionary theory on its head. Their basic argument is that rather than modeling aggression or conflict, biological systems model synergy or cooperation. Species do not fight for the same niche in an environment; they evolve to fit a free space so that they can *prevent* conflict with another species. Intrapersonally, "higher" more evolutionarily recent brain centers (like the cortex and prefrontal cortex) don't so much *control* instinct or "lower" centers, but instead provide clever ways to *help* lower centers reach their goals, leading even further to a sense of community and mutual goal setting between the organism and those around it. The human immune system in this model is more than an army which attacks invaders; it is a sense of wholeness, of mind, body and emotional well being. In the new biology, opposing parts or individuals seem meant to mate and appreciate each other, not fight.

How is this like everyday cognitive model development? For one thing, although biological entities are clearly individuals, they are also part of a larger whole, just as a cognitive event is part of a context. In this theory the whole doesn't subsume us or make us unimportant; rather it desperately needs our specialness to reach its own goals. The context or species needs us. The part or individual provides the means; the whole provides a large part of the motivation and meaning. Empty evolutionary niches go quickly out of ecological balance if they've lost touch with their meaning. Immune systems having missing elements turn on the very body that sustains them. Over and over this seems to be the biological message: Each part is important; each part is related; each part *becomes* through relationship, in context, and only in context.

What does this theory suggest for cognition? Meacham and Emont (1989) have noted that most problem solving is social, not individual. Johnson and Johnson (1975) and collaborators have demonstrated the value of cooperative learning and group work, but school systems seem slow to hear them. Kohn (1987) has listed the detrimental effects of competition in schools or workplaces, but getting help from a knowledgeable person to solve a problem in those settings is considered cheating and "bad," to date. The suggestion offered by the new biology seems to be to maximize individual cognitive growth by capitalizing on belonging to a larger whole and working together.

Chaos Theory in Mathematics

Chaos theory is a new mathematical model used to describe phenomena as different as weather, the structure of coastlines, brainwave patterns, normal or abnormal heartbeat patterns, and the behavior of the mentally ill (Alper, 1989; Cavanaugh, 1989; Crutchfield, Farmer, Packard, & Shaw, 1986; Gleick, 1987; Pool, 1989). Chaos theory answers questions about the orderly and flexible nature of apparent disorder by describing dynamic complex systems with non-linear equations. One striking feature of chaotic systems is the way in which a tiny perturbation can lead to complete reordering of the entire pattern of the system (termed "the butterfly effect"). Another is the way a seemingly random set of events, after many iterations, can coalesce around a point in an apparently orderly way so as to give the impression of a dominant feature of some sort analogous to a dominant personality trait being present (termed "strange attractors"). Think of chaos as organized disorder (as opposed to sheer randomness, or *dis*organized disorder). Without some chaotic flexibility, some orderly readiness to fluctuate built into the system, a system (especially one like the heart or the brain) is too rigid to adapt and live. For example, a rigid heartbeat pattern (no chaos) can't effectively correct for a small perturbing error, so a heart attack occurs; a rigid brain wave pattern can't respond to an intellectual challenge, so poor performance results. Chaotic disorder is *non*random and is a kind of potential to correct errors. Chaos is an order enfolded into apparent disorder, the pattern in the hologram, akin to the "implicate order" described by Prigogene and Stengers (1984). But it makes even a very minor event powerful enough to create major effects. Chaos theory gives a rationale for synchronous effects since it demonstrates entrainment in which one system locks on to the mode and pattern of another nearby system so even the minor event in one system can move other systems too.

What might such a theory imply in general? First, it suggests that there is more than one sort of disorder. Useful (chaotic) disorder provides fresh options and room to maneuver to correct for past errors; useless disorder provides nothing. Second, it suggests the immense importance of each element in the system for the final outcome. As computer models demonstrate, the perturbation caused by the butterfly's wing can alter the weather pattern! Third, it suggests the importance of openness to innovation to provide natural sorts of corrective devices during complex events.

What does this model suggest for everyday cognitive theory? How are these two models linked? First, if there are two types of disorder—useful and useless—in systems, we as therapists or problem solvers must learn to foster creative disorder or a larger problem space to foster cognitive development. This will lead to greater adaptive flexibility for the thinker. Second, chaos theory points to the potentially tremendous importance of each person's contribution during the information exchange process. One individual can totally alter everyday interpersonal cognitive meaning or dynamics, the way one shift in wind can alter a complete weather pattern. In the rapidly changing everyday cognitive situation, one agent can have real and far-reaching power to influence change and cognitive meaning. These are just two similarities between the models.

New Cognitive Science Models

One new key area now developing within cognitive science is called postformal thought (Commons, Richards, & Armon, 1984; Commons, Sinnott, Richards, & Armon, 1989; Sinnott, 1989c; Sinnott & Cavanaugh, 1991). It includes cognitive epistemology (or the knowing of reality) and lifespan development. This cognitive development is theorized to be accompanied by increases in social-cognitive experience and skills. Postformal Piagetian thought is one theory describing this development (Sinnott, 1984). Such cognitive approaches go beyond traditional information processing approaches. Postformal thought is a complex way of solving problems which develops with social experience, usually in mature adulthood. It allows a person to solve problems even in situations where many conflicted belief systems and priorities overlap. In postformal thought in Sinnott's version of the model the solver faces multiple conflicting ideas about what is true. She or he realizes that it's not possible to get outside the mind to find out which "truth" is "TRUE," but that a solution must be found to the problem anyhow. She or he then realizes that the truth system she or he chooses as true will become true, especially in relation to other people, as she or he lives it to a conclusion, within the limits of other active knowers in the system.

The main characteristics of these relativistic postformal cognitive operations (Sinnott, 1984) are: (a) self reference; and (b) the ordering of formal operations. Self reference is a general term for the ideas inherent in the new physics (Wolf, 1981) and alluded to by Hofstadter (1979) using the terms *self-referential games, jumping out of the system,* and *strange loops*. The essential notions are that we can never be completely free of the built-in limits of our system of knowing, and that we come to know that this very fact is true. This means that we take into account, in all our decisions about truth, the fact that all knowledge has a subjective choice component and therefore is, of necessity, incomplete. So, any logic we use is self-referential logic. Yet we must act, and do so by making a lower-level decision about the higher level rules of the game (nature of truth), then play the game by those rules. Sooner or later we come to realize that this is what we're doing. We then can consciously use self-referential thought.

The second characteristic of postformal operations is the ordering of Piagetian formal operations. The higher level postformal system of self-referential truth decisions gives order to lower level formal truth systems and logic systems, one of which is somewhat subjectively chosen and imposed on data.

Now this is the logic of the new physics (relativity theory and quantum mechanics) (Sinnott, 1981). New physics is the next step beyond Newtonian physics and is built on the logic of self-reference. It's reasonable that the development of logical processes themselves would follow that same progression to increasing complexity. Some characteristics that separate new physics thinking from earlier forms can be found in Table 5.1.

A new type of cognitive coordination forced by everyday and social cognitive task demands occurs at the postformal level. Another kind of coordination of perspectives also seems to happen on an emotional level, over developmental time (Labouvie-Vief, 1987). This coordination parallels the cognitive one, and probably is in a circular interaction with it. It is expected that postformal thought will be adaptive in a social situation with emotional and social components (Sinnott, 1984) because it is hypothesized to ease communication, to reduce information overload, and to permit greater flexibility and creativity of thought. The postformal thinker knows she or he is helping create the eventual truth of a social interaction by being a participant in it and choosing to hold a certain view of the truth of it.

Postformal thought has an impact on one's view of self, the world, other persons, change over time, and our connections with one another over time (Sinnott, 1981, 1984, 1989c, 1991b). It represents the way one knows or understands ideas such as those in the new sciences, that is, in all the models we've examined so far in this chapter.

Is postformal cognitive development related to everyday cognition? In earlier work, Lee (1987, 1991) has discussed some of the points of impact of this theory on the teaching process. Johnson (1991) has examined postformal thought as it relates to the learning process in international development programs. One conclusion we can draw is that effective master teachers and change agents show characteristics of postformal thought and complex cognitive processes. They can bridge across belief systems, entertain several views of truth and work well in complex social realities. They can create the necessary chaos for the flexible change that must accompany learning. They can create a cooperative learning environment in which a dialogue between "teacher" and "learner" takes place, thereby honoring the truth of both. And by both modeling such thought and by challenging any Newtonian/conflict-based/inflexible world views of their students, they provide the best conditions for the development of postformal thought in their students. This is one example of the impact of everyday cognitive processes.

The argument or discussion here appears somewhat circular, of necessity. We think in a certain way, then become aware of those processes in our everyday thought. This is another case of the paradox of the conscious mind reflecting on itself and seeing itself through its own filters (Hofstadter, 1979).

In Table 5.3 is a summary of some key principles of the new sciences with one example of an everyday cognition analog for each.

TABLE 5.3
Summary Table: Sample of Potential Everyday Cognition Analogs of New Sciences Principles

Concept	*Implication*	*Everyday Cognition Analog*
Physics		
Relativity, quantum physics: Multiple contradictory realities can all be true depending on vantage points.	Multiple truths are "true" depending on one's vantage point and choice.	Bridge between equally valid problem solving realities; two persons create mutually contexted realities.
Biology		
Cooperative evolution: Species don't fight for the same niche; they evolve to fit interrelated, nonoverlapping niches.	Individual organisms are important as individuals and as somewhat irreplaceable members of a whole system.	Tasks need context to give meaning; individuals cocreate social realities.
Mathematics		
Chaos theory systems have adaptive disorder by design, which leads to unpredictable outcomes.	Some disorder is nonrandom and is adaptive. Systems that tolerate disorder survive. Each system has an impact on the whole.	Fostering creative disorder in the learning process; individual's input matters tremendously in life-span cognitive development.
Cognitive Science		
Lifespan cognitive development and postformal Piagetian thought. Mature knowers partly choose and construct TRUTH, especially social truth.	It characterizes the thinking of mature adults. This thinking is based on a self-referential view of truth and is a new physics/new biology/chaos model of knowing reality.	Master teachers and good social change agents are postformal; persons who are adaptive in everyday setting are complex thinkers.

ANSWERS TO THE BETTER QUESTIONS CAN ADD TO THEORIES OF MIDLIFE AND LATER LIFESPAN DEVELOPMENT

We can use the new general questions generated by everyday cognitive studies and the new sciences to stimulate our thinking about models of the course of middle and later lifespan development. If we are studying everyday cognition in mature adults we also are almost compelled to analyze midlife development as a key to understanding cognitive context, meaning, and problem solving factors. Midlife questions *give context and form* to those variables which impact on cognitive processes in adults. As seen in Fig. 5.1, midlife issues also *motivate* thinking about the shifting nature of reality in everyday cognitive events. The psychologist's awareness that the person has such thought processes helps him or her explain the behavior of the person at that point in life.

Several theories of midlife development already exist, although few involve a significant cognitive component. But if, as General Systems Theory (Sinnott, 1989c) suggests, we organisms are multilevel nested systems, a cognitive component is always present during development. If, as new physics suggests, we are cocreators of reality by our knowing of it, by our use of cognitive components we may define the reality of our own lives. If the new biology is correct that we have evolved as complementary, cooperative species, our cognition would be joint with other organisms. If the forms and functions derived from chaos theory hold true, we see the tremendous potential that one small action (cognitive or otherwise) can have for differentiation of an organism over time. Cognitive factors would seem to have tremendous potential for shaping not only everyday behavior, as cognitive therapists have pointed out, but also for shaping everyday life trajectories as realities about identity are created over time. We search for understanding of reality because of midlife issues; we understand our changing world as we come to understand ourselves changing over time. Perhaps we understand our changing world in terms of our own changing and understand our changes in light of what we see as changing reality around us (Sinnott, 1989c, 1992a, 1992b, in press-a,b,c,d).

In thinking about this further, let's examine one or two typical tasks of midlife and old age that are present in various cultures or theories. At the entry point of midlife, perhaps around the age of 30, the younger adult must make a choice of a way to go, of a life to choose, in industrial cultures of multiple possibilities (Perry, 1975; Levinson, 1978). Even as she or he sees the relativity of many truths she or he must make a passionate commitment (Frankl, 1963; Perry, 1975; Polanyi, 1971) to live out one choice or truth. That choice also involves relinquishing several illusions including (Gould, 1978) that there is only one correct way to procede in life, and that one's parents have the knowledge of that single way.

Erikson (1950) describes the tasks of the midlife and old individual as developing generativity and integrity, that is, a mentoring and caring for others, a creation of persons and contributions which will outlast the self, and a sense of the satisfying completeness of one's life story and one's place in the overall story of life. Again the sense of meaning is in relations with others and the creation of a personal truth. Many authors speak of midlife as a time to deepen commitment and to choose deliberately what one's life will mean (e.g., Havighurst, 1953). One must choose when (and why!) to deploy one's resources, newly aware of their limits. This choice of meaning, if it is truly adaptive, also incorporates one's emotional side, allowing for conscious orchestration of emotional and cognitive life (Labouvie-Vief, 1987) in a way that leads (we hope) to maturity and wisdom.

The midlife adult begins to see a bigger picture that involves time and persons existing before and after him or her. As Riegel (1975) suggests, discord, dissatisfaction or disharmony, whether from other people, completed accomplishments, a rapidly shortening lifetime, or the pressure of multiple social roles, demands a new adaptive stance. For example, Jung (1930/1971) speaks of the new incorporation of the dark side of the personality into the conscious self, at midlife.

The main tasks in these theories and observations involve bridging realities, entering the reality of another person, and developing complex concepts of the self, of success, of personal continuity. By this time in life the person has gathered the skills and the experience to make this potential midlife leap in thinking structures. Spurred by everyday social encounters, fresh from the everyday problem solving tasks of creating a marriage, a long term friendship, a parent-child relationship, an organization, a social role, a self, the adaptive midlife adult is primed to make new realities.

Like the developing child in Piagetian theory, the midlife adult seems to develop using assimilation and accommodation to reach new ways of filtering life with a new postformal logic. We would be misguided to try to test a child for Piagetian formal operations by giving that child a concrete operations conservation problem. Sure, the child might just find a way to talk in formal operations terms in spite of task limitations, but it's not likely. We'd get better results by giving a formal operations problem. In the same vein, everyday cognitive tasks may be a better way than lab tasks alone to illumine the basic cognitive aspects of midlife development and the practical cognitive parameters of successful adaptation in that stage.

ILLUSTRATION: SINNOTT'S EVERYDAY COGNITIVE STUDIES OF PROBLEM SOLVING

Since 1974, I have been working on studies of lifespan everyday problem solving and memory based, for the most part, on Piaget's model of logical development, on artificial intelligence, and on concept problem-solving literature. These studies have led to results that were intriguing enough that I eventually created my own theory of postformal logical development during midlife and old age (Sinnott, 1984). From the start I used formal problems drawn from Piaget's work, and created increasing degrees of familiarity in the contexts of those problems. The context is more or less social (interpersonal). Thus the problems have a more or less familiar context, although I can't say they are everyday in the sense of routine for each participant. The problems are given, so far, in interview format, although a computerized version is near completion. The computerized version is a series of problems that yields both "postformal" and "cognitive style" scores useful for estimating cognitive complexity of teachers, trainers, therapists, etc. Some of the results discussed later came from intense analyses of individuals thinking aloud, and some of these results lend themselves to individual differences or style difference analyses. Other results are nomothetic, addressing the average performance of a group. Some results of these studies have been published elsewhere (e.g., Sinnott, 1984, 1989c, 1989d, 1989f, 1989g, 1991b) and can be applied to many fields and problems (Sinnott, in press-b, in preparation).

My first goal in the research in 1973 or so was to see what mature adults *were* doing when they seemed to fail Piagetian problem solving tasks. I wanted to describe their processes, make sense of them, and see how age influenced them. What they seemed to be doing seemed both more skilled and less skilled than what young adults did. It seemed much more adaptive than what young adults did. When I asked the question about basic cognitive problem solving processes in adaptive cognition, I was forever more a prisoner of everyday cognitive science. I had been kidnapped by an army of more intriguing questions.

My studies produced descriptions of basic processes in adaptive complex everyday problem solving and memory, and mechanisms underlying individual differences. They addressed parameters for performance at a complex (rather than elementary) level, and involved experiments to induce those achievements. They explored some social/interpersonal elements which led to, or were occasions of, complex thought that was adaptive or functionally useful. They examined case histories. They led to paradigms for studies for which testing materials could range from very abstract (e.g., a paper-and-pencil "rat testing maze," based on a real building floor plan), through more realistic variations (e.g., a video of real floor plan), to realistic variations (e.g., walking a participant around the real floor of the building), the object being to experimentally control every major factor but the naturalness of stimulus materials (which was experimentally manipulated on a continuum).

As an example of everyday cognitive studies, a summary of what I consider some of the more interesting parts of my everyday cognitive research is in Table 5.4. The Table also contains a list of mechanisms addressed in these studies. Space precludes describing each one here. Figure 5.2, reprinted from Sinnott and Cavanaugh (1991), gives an example of a simple model of processes encountered in solving a cognitive problem viewed as realistic by the solver. More than 200 protocols of individuals solving problems they saw as realistic or everyday are on file in my Center; a similar set from well or poorly adjusted couples solving problems is being collected.

Examination of those studies will demonstrate that we have made a start. Development of basic *everyday* cognition mechanisms seems to reliably occur, can be the object of experimental manipulation, can be modeled, and is different from (but related to) information processing, artificial intelligence, or concept formation mechanisms as usually described in the literature. Asking broader questions—those related to functional cognition—is harder, but full of surprises and new information. Isn't this what science is about, new understanding? The protocols are full of emotion, intention, other people, daydreams, personal history, expressions of personal values and attitudes. These have been absent from cognitive studies for too long.

TABLE 5.4
Some Sinnott Everyday Cognition Studies, and Some Cognitive Mechanisms Addressed Within Them

Reference	*Topic*	*Mechanism*
Sinnott (1975)	Logical problem solving of middle aged and older *Ss*	Some types of problem contexts (i.e., every-day vs. abstract) enhance logical performance and attenuate cross-sectional age effects.
Chap and Sinnott (1977-1978)	Concrete and formal Piagetian tasks by older *Ss* in institutions and community	Logical performance seems heavily influenced by *Ss'* setting.
Sinnott and Guttman (1978a)	Logical abilities and everyday problem solving in older *Ss*	Logical abilities only explain part of success or failure on every day problems.
Sinnott and Guttman (1978b)	Decision making of older adults	Everyday decision making has special elements of Piagetian logical operations and dialectical logic, elements that predict decision outcome.
Sinnott (1984)	Reasoning, problem solving, lifespan sample, think aloud methods	We can list cognitive operations peculiar to everyday complex problem solving. Complex thought is a constructed reality that limits resources needed to process information.
Cavanaugh, Kramer, Sinnott, Camp, and Markley (1985)	Problem solving is middle aged and older adults	Mature adults partly construct their everyday logical reality, based on their experiences constructing social relations.
Sinnott (1985)	Problem solving of singles and dyads	Partners solved problems logically more often than singles, partners' consensus favored abstract logical answers.
Sinnott (1986a)	Forgetting rate; lifespan sample	Forgetting rate is equivalent for younger and older *Ss* for equally practiced everyday tasks.
Sinnott (1986b)	Prospective memory over time in middle aged, younger, older *Ss*	Memory declines over time and poorer memory in older *Ss* is found for incidental memory but not for motivationally salient prospective memory.
Sinnott (1987e)	Spatial memory, lifespan sample	A paradigm is described to experimentally test memory for space in more or less realistic contexts, controlling for other factors.
Sinnott (1987c)	Activity memory, lifespan sample	Everyday memory for action is more motivating than incidental (I) but less motivating than prospective (P) memory. So, it is remembered less than P but more than I. Intrinsic context-related motivation influencing performance even within everyday tasks.
Sinnott (1987b; 1988)	Problem solving lifespan sample	"noncognitive" processes such as emotion influence problem solving and can improve models of that process.
Sinnott (1989e)	Problem solving	Future research tasks and pitfalls are listed. This chapter offers a rudimentary research plan for studies of everyday problem solving.

(*Continued*)

TABLE 5.4
(Continued)

Reference	*Topic*	*Mechanism*
Sinnott (1989a, 1989b, 1989g, 1989h)	Individual and group differences in complex problem solving and reasoning over lifespan	Models of everyday problem solving can be constructed, reliably. Everyday operations have information processing skill correlates and are influenced by goal clarity and heuristic availability. Everyday memory and everyday problem solving operations are related.
Sinnott (1989f)	Memory, lifespan	General systems theory provides a model for further everyday memory studies.
Sinnott (1989c)	Cognitive change processes	General systems theory provides a model for process studies.
Sinnott (1987a); Sinnott, Plotz, and Robinson (1990)	Health (blood pressure) and cognitive function over 12 yrs; lifespan samples	Thinking is influenced in complex ways by physical factors at various ages. Seven models are tested. Everyday health matters to cognition.
Sinnot, Bochenek, Walters, and Klein (1990)	Spatial memory	Encoding modality has differential impact on everyday vs. abstract performance on same task.
Sinnott (1991b)	Problem solving in clinical context	A model provides mechanisms for describing the process partly by including the important "noncognitive" variables in the process.
Sinnott (1991a)	Problem solving, lifespan sample	An experiment shows the impact of manipulating "noncognitive" factors on logical problem solving performance.
Sinnott, Bochenek, Kim, Klein, Robie, Dishman, and Dunmyer (1992)	Spatial memory, older and younger *Ss*	Mechanisms of encoding, retrieval differentially influence old and young on everyday or abstract maze versions.

We in cognitive science need to ask ourselves if we haven't (metaphorically speaking) spent too many years in the great land of Oz, seeking the Wizard of objectivity, thinking we cannot have studies that incorporate everyday basic human situations, and human elements like hearts. The Wizard might be largely illusion. The new physics suggests that he is. We might already be on a good path with a heart and a brain, studying fully human everyday thought, if we choose to recognize it. This path, as don Juan could have said, can potentially make for a joyful, useful journey in science. We might decide there's no place quite like home for understanding the mind.

ACKNOWLEDGMENTS

This chapter, which was originally an invited presentation at the Twelfth West Virginia University Lifespan Development Conference, Morgantown, WV,

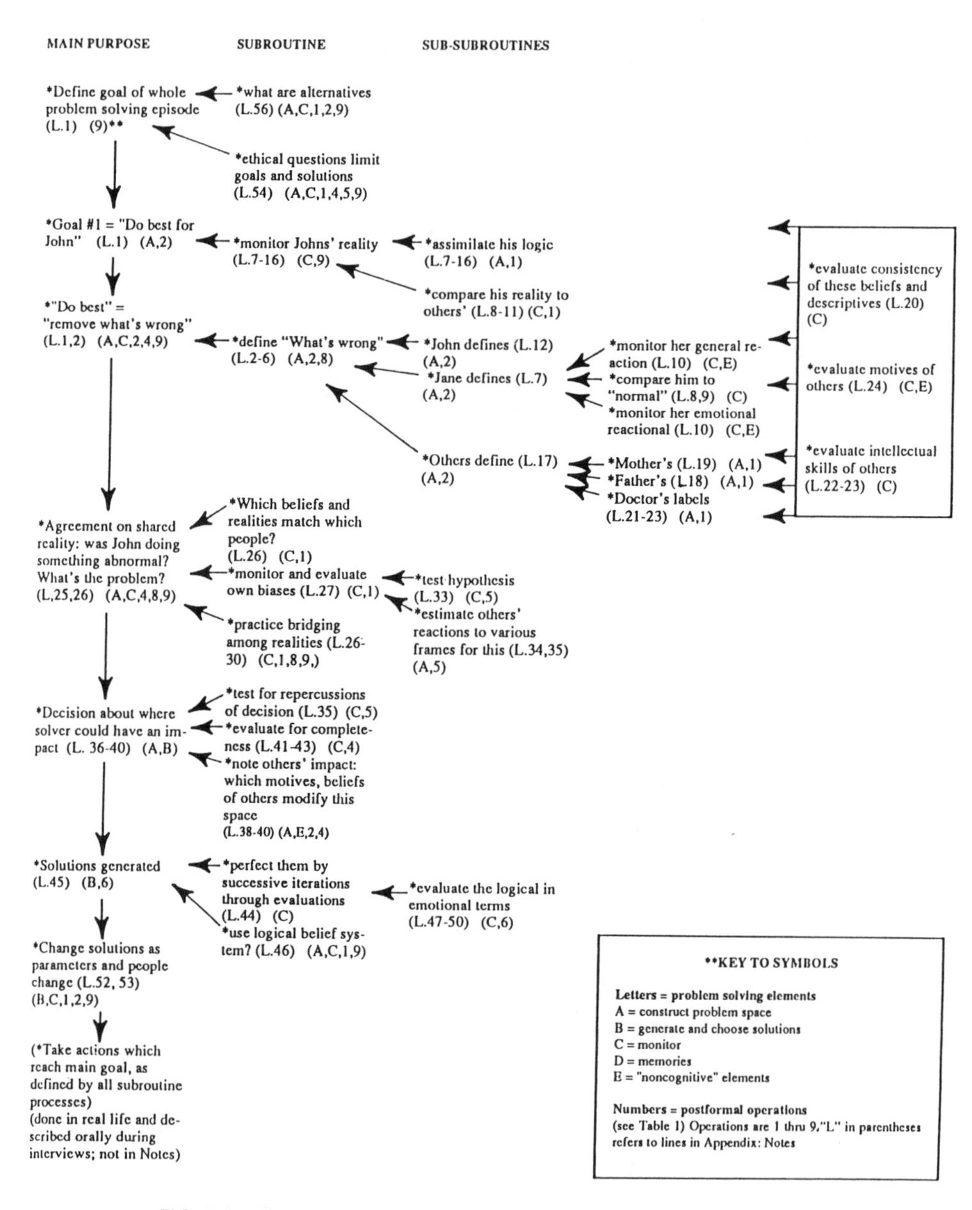

FIG. 5.2. Summary of processes used by key informant based on her notes alone. From Sinnott (1991b). In *Bridging Paradigms: Positive Development in Adulthood and Cognitive Aging* by J. Sinnott & J. Cavanaugh (Eds.). Copyright (1991). New York: Praeger. An imprint of Greenwood Publishing Group, Inc., Westport, CT. Reprinted by permission.

1990, was prepared with support from a Faculty Research Grant from Towson State University. Portions of the projects reported here were supported by grants from the Public Health Service, from NIH, and from Towson State University. The cooperation of the Gerontology Research Center, NIA, NIH, and its Longitudinal Study of Aging are gratefully acknowledged.

REFERENCES

Alper, J. (1989). The chaotic brain: New models of behavior. *Psychology Today. 23,* 21.

Augros, R., & Stanciu, G. (1987). *The new biology.* Boston: New Science Library.

Bronowski, J. (1976). *The ascent of man.* Boston: Little, Brown.

Campbell, J. (1988). *The power of myth.* New York: Doubleday.

Castaneda, C. (1981). *A separate reality.* New York: Washington Square Press.

Cavanaugh, J. (1989). *Utility of concepts in chaos theory for psychological theory and research.* Paper presented at the Fourth Adult Development Conference at Harvard University, Cambridge, MA.

Cavanaugh, J., Kramer, D., Sinnott, J. D., Camp, C., & Markley, R. P. (1985). On missing links and such: Interfaces between cognitive research and everyday problem solving. *Human Development, 28,* 146–168.

Chap, J., & Sinnott, J. D. (1977–1978). Performance of institutionalized and community-active old persons on concrete and formal Piagetian tasks. *International Journal of Aging and Human Development, 8,* 269–278.

Commons, M., Richards, F., & Armon, C. (Eds.). (1984). *Beyond formal operations.* New York: Praeger.

Commons, M., Sinnott, J. D., Richards, R., & Armon, C. (Eds.). (1989). *Adult development II: Comparisons and applications of adolescent and adult developmental models.* New York: Praeger.

Crutchfield, J. P., Farmer, J. D., Packard, N. H., & Shaw, R. S. (1986). Chaos. *Scientific American, 255,* 46–57.

Erikson, E. (1950). *Childhood and society.* New York: Norton.

Frankl, V. (1963). *Man's search for meaning.* New York: Washington Square Press.

Gleick, J. (1987). *Chaos: Making a new science.* New York: Penguin Books.

Gould, R. (1978). *Transformation.* New York: Simon & Schuster.

Havighurst, R. (1953). *Human development and education.* New York: Longmans.

Hofstadter, D. R. (1979). *Gödel, Escher and Bach: An eternal golden braid.* New York: Basic Books.

Johnson, D., & Johnson, R. (1975). *Learning together and alone.* Englewood Cliffs, NJ: Prentice Hall.

Johnson, L. (1991). Postformal reasoning facilitates behavioral change: A case study of an international development project. In J. D. Sinnott & J. Cavanaugh (Eds.), *Bridging paradigms: Positive development in adulthood and cognitive aging.* New York: Praeger.

Jung, C. (1971). The stages of life. In J. Campbell (Ed.), *The portable Jung.* New York: Viking. (Originally published 1930).

Kohn, A. (1987). *No contest.* Boston: Houghton Mifflin.

Labouvie-Vief, G. (1987). *Speaking about feelings: Symbolization and self regulation through the lifespan.* Paper presented at the Third Beyond Formal Operations Symposium at Harvard, Cambridge, MA.

Lee, D. (1987). *Relativistic operations: A framework for conceptualizing teachers' problem solving.* Paper presented at the Third Beyond Formal Operations Symposium at Harvard, Cambridge, MA.

Lee, D. (1991). Relativistic operations: A framework for conceptualizing teachers' problem solving. In J. D. Sinnott & J. Cavanaugh (Eds.), *Bridging paradigms: Positive development in adulthood and cognitive aging.* New York: Praeger.

Levinson, D. (1978). *The seasons of a man's life.* New York: Knopf.

Maturana, H., & Varela, F. (1988). *The tree of knowledge.* Boston: New Science Library.

McLean, P. (1988). *Evolutionary biology.* Paper presented at the Gerontology Research Center, National Institute on Aging, NIH, Baltimore, MD.

Meacham, J., & Emont, N. C. (1989). The interpersonal basis of everyday problem solving: In J. D. Sinnott (Ed.), *Everyday problem solving Theory and application* (pp. 7–23). New York: Praeger.

Perry, W. B. (1975). *Forms of intellectual and ethical development in the college years: A scheme.* New York: Holt, Rinehart, & Winston.

Polanyi, M. (1971). *Personal knowledge.* Chicago: University of Chicago Press.

Pool, R. (1989). Is it healthy to be chaotic? *Science, 243,* 604–607.

Prigogene, I., & Stengers, I. (1984). *Order out of chaos.* New York: Bantam.

Riegel, K. (1975). Adult life crises: A dialectical interpretation of development. In N. Datan & L. Ginsberg (Eds.), *Lifespan developmental psychology: Normative life crises* (pp. 99–129). New York: Academic Press.

Sinnott, J. D. (1975). Everyday thinking and Piagetian operativity in adults. *Human Development, 18,* 430–444.

Sinnott, J. D. (1981). The theory of relativity: A metatheory for development? *Human Development, 24,* 293–311.

Sinnott, J. D. (1984). Postformal reasoning: The relativistic stago. In M. Commons, F. Richards, & C. Armon (Eds.), *Beyond formal operations* (pp. 298–325). New York: Praeger.

Sinnott, J. D. (1985). *Effect on goal clarity and presence of a partner on problem solving performance.* (Available from Psychology Department, Towson State University, Baltimore, Md.)

Sinnott, J. D. (1986a). *Everyday memory: With equal practice does age influence forgetting rate in a two year period?* Paper presented at the Gerontological Society, Chicago. (ERIC, ED 276911, RIE May 87).

Sinnott, J. D. (1986b). Prospective/intentional everyday memory: Effects of age and passage of time. *Psychology and Aging, 1,* 110–116.

Sinnott, J. D. (1986c). Social cognition: The construction of self-referential truth? *Educational Gerontology, 12,* 337–340.

Sinnott, J. D. (1987a). *Blood pressure/cognition relations: Alternative models of change over time.* Paper presented at American Psychological Association, New York.

Sinnott, J. D. (1987b). *Experimental studies of relativistic self-referential postformal thought: The roles of emotion, intention, attention, memory and health in adaptive adult cognition.* Invited paper presented at the Third Harvard University Conference on Positive Adult Intellectual Development, Cambridge, MA.

Sinnott, J. D. (1987c). *Memory for action taken.* Paper presented at Eastern Psychological Association, Arlington, VA.

Sinnott, J. D. (1987d). *Models of everyday problem solving: Under what conditions might older adults show postformal thought?* Paper presented at Gerontological Society of America, Washington D.C.

Sinnott, J. D. (1987e). *Spatial memory and aging: Models linking lab and life, rat and human maze learning.* Paper presented at the Cognitive Aging Conference, Atlanta, GA.

Sinnott, J. D. (1988). *"Noncognitive" processes in problem solving: Are there age differences in use of emotion, personal history, self evaluation and social factors during combinatorial problem solving?* Paper presented at the Second Cognitive Aging Conference, Atlanta, GA.

Sinnott, J. D. (1989a). A model for the solution of illstructured problems: Implications for everyday and abstract problem solving. In J. D. Sinnott (Ed.), *Everyday problem solving: Theory and applications* (pp. 72–99). New York: Praeger.

Sinnott, J. D. (1989b). Adult differences in use of postformal operations. In M. Commons, J. D. Sinnott, F. Richards, & C. Armon (Eds.), *Beyond formal operations II* (pp. 239–278). New York: Praeger.

Sinnott, J. D. (1989c). Changing the known, knowing the changing. In D. Kramer & M. Bopp (Eds.), *Transformation in clinical and developmental psychology* (pp. 51–69). New York: Springer.

Sinnott, J. D. (Ed.). (1989d). *Everyday problem solving: Theory and application.* New York: Praeger.

Sinnott, J. D. (1989e). Summary: Issues and directions in everyday problem solving research. In J. D. Sinnott (Ed.), *Everyday problem solving: Theory and applications* (pp. 300–306) New York: Praeger.

Sinnott, J. D. (1989f). General systems theory: A rationale for the study of everyday memory. In L. Poon, D. Rubin, & B. Wilson (Eds.), *Everyday cognition in adulthood and old age* (pp. 59–70). New York: Cambridge University Press.

Sinnott, J. D. (1989g). Lifespan relativistic postformal thought: Methods and data from everyday problem solving studies. In M. Commons, J. Sinnott, F. Richards, & C. Armon (Eds.), *Adult development: Comparisons and applications of developmental models* (pp. 239–278). New York: Praeger.

Sinnott, J. D. (1989h). Prospective memory and aging: Memory as adaptive action. In L. Poon, D. Rubin, & B. Wilson (Eds.), *Everyday cognition in adulthood and old age* (pp. 352–369). New York: Cambridge University Press.

Sinnott, J. D. (1991a). "Noncognitive" processes in problem solving: Are there age differences in use of emotion, personal history, self evaluation or social factors? In J. D. Sinnott & J. Cavanaugh (Eds.), *Bridging paradigms: Positive development in adulthood and cognitive aging.* New York: Praeger.

Sinnott, J. D. (1991b). What do we do to help John? A case study of postformal problem solving in a family making decisions about an acutely psychotic member. In J. D. Sinnott & J. Cavanaugh (Eds.), *Bridging paradigms: Positive development in adulthood and cognitive aging.* New York: Praeger.

Sinnott, J. D. (1992a). *Development and yearning: Cognitive aspects of spiritual development.* Paper presented at the American Psychological Association Convention, Washington, D.C.

Sinnott, J. D. (1992b). The use of complex thought and solving intragroup conflicts: Means to conscious adult development in the workplace. In J. Demick and P. Miller (Eds.), *Development in the workplace* (pp. 155–175). Hillsdale, NJ: Lawrence Erlbaum Associates.

Sinnott, J. D. (in press-a). Creativity and postformal thought: Why the last stage is the creative stage. In C. Adams-Price (Ed.), *Creativity and aging: Theoretical and empirical approaches.* NY: Springer.

Sinnott, J. D. (Ed.) (in press-b). *Interdisciplinary handbook of adult lifespan learning* (pp. 155–175). CT: Greenwood.

Sinnott, J. D. (in press-c). Postformal thought, adult learning, and lifespan development: How are they related? In J. D. Sinnott (Ed.), *Interdisciplinary handbook of adult lifespan learning.* CT: Greenwood.

Sinnott, J. D. (in press-d). Teaching in a chaotic new physics world: Teaching as a dialogue with reality. In P. Kahaney, J. Janangelo and L. A. M. Perry (Eds.), *Theoretical and critical perspectives on teacher change.* New Jersey: Ablex Press.

Sinnott, J. D. (in preparation). *Interpersonal applications of cognitive growth in adulthood.*

Sinnott, J. D., Bochenek, K., Kim, M., Klein, L., Walters, C., Dishman, K., & Dunmyer, C. (1992). Aging and memory for spatial relations. In R. L. West & J. D. Sinnott (Eds.). *Everyday memory and aging: Current research and methodology.* New York: Springer-Verlag.

Sinnott, J. D., Bochenek, K. R., Walters, C. R., & Klein, L. (1990). *Spatial memory and visual and kinesthetic coding effects*. Paper presented at Cognitive Aging Conference, Atlanta, GA.
Sinnott, J. D., & Cavanaugh, J. (Eds.). (1991). *Bridging paradigms: Positive development in adulthood and cognitive aging*. New York: Praeger.
Sinnott, J. D., & Guttman, D. (1987a). Piagetian logical abilities and older adults' abilities to solve everyday problems. *Human Development, 21,* 327–333.
Sinnott, J. D., & Guttman, D. (1978b). The dialectics of decision making in older adults. *Human Development, 12,* 190–200.
Sinnott, J. D., Plotz, J. B., & Robinson, S. (1990). Models of physiology/cognition relations: Their prevalence in literature and their utility for examining the effect of blood pressure on vocabulary and memory for designs. (*ERIC*, C. 6022595).
Wolf, F. A. (1981). *Taking the quantum leap.* New York: Harper and Row.
Wolfe, T. (1942). *You can't go home again.* New York: Sun Dial Press.

6 Everyday Reasoning and the Revision of Belief

Michael Chapman
University of British Columbia

The concept of "everyday cognition" is one of those indeterminate concepts that one must try to define before one can say very much about it. Part of the problem is that the concept is implicitly defined by contrast with "formal" or "academic" cognition. Everyday cognition refers to forms of thinking that are *not* included within the boundaries of formal or academic cognition. The problem is to say something affirmative about a concept that, in the first instance, is negatively defined. The problem of definition leads to a dilemma involving representativeness and rationality. On the one hand, everyday cognition is believed to be worth studying because it is more representative of human thought in general than formal cognition. On the other hand, formal thought and formal logic in particular have been considered for centuries as prototypical for human rationality. The dilemma consists in the necessity of choosing between the propositions (a) that much of everyday cognition is irrational to the extent that it deviates from formal thinking, or (b) that formal thought and formal logic are not the best or only standards of human rationality.

The study of everyday reasoning leads to the conclusion that one should grasp the second horn of this dilemma: that human rationality is not epitomized by formal thought in general and that studying everyday reasoning opens up a broader conception of rationality. A preliminary step in this argument is defining formal reasoning and everyday reasoning and distinguishing them from each other.

FORMAL AND EVERYDAY REASONING

In an article entitled, "Approaches to Studying Formal and Everyday Reasoning," Galotti (1989) defined reasoning in general as "mental activity that consists of transforming given information (called the set of premises) in order to reach conclusions" (p. 333). Given this definition, she went on to describe *formal reasoning* in terms of the following sets of characteristics (p. 335):

1. All premises are supplied explicitly.
2. Problems tend to be self-contained.
3. Typically, there is only one correct answer to each problem.
4. Established methods of inferences are often available.
5. The solution to a given problem can be recognized unambiguously.
6. The content of the problem generally is of limited, "academic" interest.
7. Problems are solved for their own sake, and not for any further ends.

In contrast, *everyday reasoning* was described by a very different set of characteristics:

1. The premises might be only implicit in the information provided, and some premises might not be supplied at all.
2. Problems are usually not self-contained.
3. Several possible answers might exist, not just one, and those answers might vary in quality.
4. Established procedures of problem-solving rarely exist.
5. It is often unclear when a given solution, even if it is recognized as the best available, is sufficient for one's purposes.
6. The content of the problem typically is relevant to one's personal interests.
7. Problems are often solved as a means to further ends, and not as ends in themselves.

Two things should be noted about this distinction: First, it focuses on *task* characteristics and not on cognitive *processes*. The question of whether formal and everyday reasoning differ with respect to cognitive processes is viewed as an empirical problem, not part of their respective definitions. Second, the distinction between formal and everyday reasoning overlaps with some previous distinctions, for example, with the distinction between *well-defined* and *ill-defined* problems (Glass, Holyoak, & Santa, 1979), or with that between *theoretical* and *practical* thinking (Scribner, 1986). Everyday reasoning is "ill-defined" to the extent that the information given, the permitted operations, or the desired end-state(s) are left relatively unspecified. It is "practical" to the extent that problems are solved as a means of achieving the goals of practical action.

However, Galotti's distinction does not wholly coincide with those previous distinctions, nor with other possible construals of "formal" and "everyday" reasoning.[1] For example, one might object to her characterization of formal reasoning as being of academic interest only or as directed to the solution of problems for their own sake. Formal reasoning can surely be put to practical or even personal ends as in the case of scientists working to develop useful technology while seeking to further their own careers. Similarly, one might object to the claim that established procedures for everyday reasoning do not exist, insofar as attempts have been made to codify the heuristics involved in everyday reasoning (e.g., the systems of "nonmonotonic reasoning" described later in this chapter).

In defense of Galotti's distinction, one might argue that formal and everyday reasoning should be considered *family-resemblance concepts* and, therefore, that a particular instance of either concept need not possess all the characteristics listed (Rosch, 1987; Wittgenstein, 1958, p. 32e). Thus, formal reasoning would not *necessarily* be of limited, academic interest, but only frequently so. And to the extent that heuristics for "everyday reasoning" have been codified, one might argue that the distinction between formal and everyday reasoning becomes blurred and that reasoning guided by such heuristics acquires, by that fact alone, a more "formal" and less "everyday" character than reasoning without the benefit of such guidance.

In this chapter, Galotti's distinction is employed for the limited purpose of framing the issues to be discussed. For that purpose, it is well suited. As Wittgenstein (1958, p. 34e) pointed out, an "indistinct" concept may be nonetheless useful for that fact. Indeed, it might be just what one needs.

Having made the distinction between formal and everyday reasoning, Galotti (1989) went on to review some relevant research. On the one hand, the development of formal reasoning appears to depend in part on the experience of formal education (Laboratory of Comparative Human Cognition, 1983). For example, it apparently does not occur spontaneously in cultures in which formal education is not the rule. On the other hand, even educated adults in Western industrialized countries often have great difficulty with some principles of formal reasoning, drawing inferences from negative premises being only one example (Matlin, 1983). Finally, the ability to solve everyday reasoning problems generally has not been found to correlate very highly with the ability to solve formal reasoning problems (Wagner & Sternberg, 1986), although some exceptions have been reported (Ceci & Liker, 1986). These research findings are significant, both because they tend to validate the distinction between formal and everyday reasoning and because they comprise a set of "core phenomena" that must be explained in any *general* theory of reasoning (i.e., in any theory that encompasses both formal and everyday reasoning).

The rest of Galotti's article was devoted to a description and evaluation of three potential candidates for such a general theory: (a) Sternberg's (1983) *com-*

---

[1] I am indebted to Hayne Reese for the following examples.

ponential approach, according to which reasoning can be analyzed into its component, real-time processes; (b) the *rules/heuristics/schemata* approach, according to which reasoning is a process of applying specific rules, heuristics, or inference schemas, usually drawn from among the theorems of propositional logic (Braine, 1978; Osherson, 1975; Rips, 1983); (c) the *mental models/search* approach, according to which the problem is represented in some problem-solving "space" and the solution is implicit in that representation (Johnson-Laird, 1983; Newell, 1981). Galotti summarized the strengths and weaknesses of each approach and concluded, quite reasonably, that each has its own strengths and weaknesses.

The foregoing theories are not reviewed in any further detail in this chapter. Instead, an argument is advanced which differs from each of them, and even from Galotti's conceptualization of the problem. The basic argument can be summarized in terms of the following propositions:

1. The everyday concept of reasoning is not restricted to intrapsychic inference, but includes (among other things) the social and communicative activity of *argumentation*.
2. Reasoning as intrapsychic inference develops in part from reasoning as argumentation through a process of *sociogenesis*.
3. Everyday reasoning can be distinguished from formal reasoning by means of their respective *sociogenetic origins*.

In the following sections, each of these claims is examined in turn.

ARGUMENTATION AS A FORM OF EVERYDAY REASONING

The claim that reasoning in the ordinary sense is not limited to intrapsychic inference is not meant to imply that the latter is not a legitimate object of investigation. The point is rather that if one limits the concept of reasoning at the outset to intrapsychic inference, one is likely to exclude activities that are highly relevant to a theory of everyday reasoning. One such activity is the attempt to convince other persons of a particular point of view by providing them with *reasons* for doing so. In ancient Greece and medieval Europe, this type of activity was known as *rhetoric* or *dialectic* (Barilli, 1989). In recent years, it has been studied under the rubric of *argumentation* (Van Eemeren, Grootendorst, & Kruiger, 1984). It is distinguished from mere persuasion by a commitment to decision making based on the weighing of reasons for and against the viewpoint in question. So understood, argumentation is by no means limited to formal debate, but includes as well everyday efforts to convince other persons of a particular belief or opinion.

One group of researchers has defined argumentation as "the social, intellectual, verbal activity serving to justify or refute an opinion . . . and being directed toward obtaining the approbation of a judge who is deemed to be reasonable" (Van Eemeren & Grootendorst, 1982; see also Van Eemeren et al., 1984). They go on to describe argumentation as a complex speech act in which the intended outcome is to convince another person to accept a new opinion (or to reject an opinion currently held) by providing that person with propositions which *justify* the opinion to be accepted (or *refute* the opinion to be rejected).

Argumentation deserves to be considered an example of *reasoning* more generally, both because it is based on an appeal to *reasons,* and because it involves inferential processes. For example, inference is involved in the effort to maintain consistency among one's assertions in the course of a discussion. The affirmation of a given proposition may commit one *by implication* to the acceptance of additional, unstated propositions. A common move in argumentation is to show that some of one's opponent's statements have implications which conflict with other propositions that the opponent accepts. The *reductio ad absurdum* is only one example of this strategy. The avoidance of inconsistency among one's commitments in argumentative discourse corresponds to the logical principle of noncontradiction, and the identification of the unstated commitments implied by a stated proposition follows inferential processes having close structural parallels in deductive logic (Mackenzie, 1989). For example, Mackenzie (1989, p. 106) discussed *modus ponens* as follows: "If Bob is committed both to 'P' and to 'If P then Q,' he is liable to an objection for inconsistency should he deny, withdraw, or challenge 'Q'." The close relation between the kinds of inferences occurring in interpersonal discourse and those sanctioned by deductive logic is also alluded to in Aristotle's definition of a syllogism as "discourse in which, certain things being stated, something other than what is stated follows of necessity from their being so" (Aristotle, *Prior Analytics* 24b, in Hutchinson, 1952, p. 39).

Despite such structural resemblances, the context of inference in interpersonal argumentation differs from that of deductive logic in some fundamental ways. For example, persons engaging in argumentation may make inferences based on assumptions about a speaker's *intentions* in affirming a given proposition. In contrast, such interpretive assumptions have no place in logical deduction, and psychologists studying reasoning as intrapsychic inference accordingly consider them as irrelevant or misleading. Logical reasoning is said to involve "analytical comprehension"—understanding a sentence for what it says in itself, rather than in terms of the intentions of the person who may have uttered it (Braine & Rumain, 1983, pp. 267–268).

If, in contrast, the concept of reasoning is expanded to include interpersonal argumentation, then such interpretive processes become part of the phenomena to be studied. Perhaps the best known attempt to summarize the interpretive assumptions involved in interpersonal communication is Grice's (1975) *maxims of conversation.* Briefly, the idea is that all discourse is a cooperative enterprise governed in part by the following norms:

1. The maxim of *quantity:* "Be informative." Provide as much information as is required to communicate your meaning (but not more information than is necessary).
2. The maxim of *quality:* "Be truthful." Avoid statements that are either false or unfounded.
3. The maxim of *relation:* "Be relevant." Speak to the point.
4. The maxim of *manner:* "Be perspicuous." Avoid ambiguity and obscurity of expression; be brief and orderly.

The underlying assumption is not that speakers necessarily adhere to such norms on all occasions, rather that successful communication depends at least on a general adherence to them. Accordingly, competent speakers acquire an expectation that those norms will be adhered to, other things being equal, by anyone seriously intending to engage in communicative interaction. Against the background of such general expectations, specific violations of the maxims of conversation can be meaningful. For example, a speaker who remarks of an acquaintance, "He is doing very well; he likes his colleagues, and he hasn't been to prison yet" (Grice, 1975), is evidently violating the maxim of relation insofar as the second half of the sentence is not obviously related to the first. However, under the assumption that the speaker is a competent language-user (e.g., that he or she has no mental disorder that results in habitually disconnected speech), a listener might conclude that this deviation from the norm was meaningful—that the speaker intended to imply something about the person mentioned (or his new occupation) without saying it directly. The listener is then likely to infer what it was that the speaker intended to communicate in this way. Grice (1975) called such forms of indirect communication *conversational implicature,* and it is often referred to more generally as the "logic of conversation."

Such oblique communications do not necessarily characterize argumentative discourse, but assumptions regarding a speaker's intentions are relevant and necessary for deciding to what beliefs or opinions speakers actually have committed themselves in asserting particular statements. The inferential process of drawing out the implications of a speaker's assertory commitments depends on assumptions regarding the intended scope of the commitment implied in making particular statements. A common defense against charges of self-contradiction in argumentation is a qualification of one's previously unqualified assertions (Rescher, 1977). Such qualification can be interpreted as a process of making one's assertory commitments more specific.

In summary, argumentation is a form of everyday reasoning, because it is based on a consideration of reasons and because it involves inference. Some forms of inference (such as the *modus ponens* rule) are shared with formal reasoning, but others (like conversational implicature) are distinctive of communicative interaction. Much everyday reasoning may occur in the context of interpersonal argumentation, and not only in the context of intrapersonal problem solving. Therefore, the study of everyday reasoning should be expanded to include the study of argumentation (Berkowitz, Oser, & Althof, 1987; Leadbeater, 1988) as it occurs in everyday situations, as well as the study of everyday problem solving.

THE CONCEPT OF SOCIOGENESIS

One might think that this appeal to an expanded concept of reasoning is just so much haggling over definitions and the division of labor. Why shouldn't cognitive psychologists continue to study intrapsychic inference and leave argumentation to rhetoricians and social psychologists? That would be no problem, if one could assume that intrapsychic inference and verbal argumentation were essentially independent. However, if intrapsychic inference develops in some sense from interpersonal argumentation, then the two would not be totally independent. The latter thesis brings us to the *second* step in the argument mentioned earlier: that intrapsychic inference develops through *sociogenesis,* a process by which certain features of mental life develop through some form of internalization or interiorization of social interaction.

So defined, sociogenesis was a feature of the theories of both Piaget and Vygotsky. For example, in *Judgment and Reasoning in the Child,* Piaget (1924/1928) described the development of logical thinking as the result of children's confrontation with other points of view and of their subsequent need to justify their own viewpoints. Indeed, he went so far as to write, "Logical reasoning is an argument which we have with ourselves, and which reproduces internally the features of a real argument" (p. 204). Vygotsky (1978) carried the concept of sociogenesis even further, arguing that all higher cognitive processes develop from social interaction. However, only the sociogenesis of reasoning is relevant in the present context. The main argument, consistent with the Piagetian thesis just stated, is that intrapsychic inference develops through the interiorization of interpersonal argumentation. A detailed description of the process by which intrapsychic forms of cognition originate from external interaction is beyond the scope of this chapter (see Furth, 1981, Reading 3 for a description of "interiorization" in Piagetian theory, and Wertsch & Stone, 1985, for a Vygotskian account of "internalization"). Instead, I focus on the implications of the process of sociogenesis for an understanding of formal and everyday reasoning.

SOCIOGENETIC ORIGINS OF FORMAL AND EVERYDAY REASONING

The idea that the process of sociogenesis is relevant for understanding the nature of reasoning is based on the developmental assumption that the character of a phenomenon is influenced by the circumstances of its origins. With this proposition, we arrive at the *third* step in the argument advanced in this chapter: Not only is the character of reasoning as intrapsychic inference influenced by its sociogenetic origins, but also the *differences* between everyday reasoning and formal reasoning can be explained in terms of differences in their *respective* sociogenetic origins.

Consider *formal reasoning* first. The main thesis is that "formal thinking" develops from argumentation through the mastery of "formal methods" (Fig. 6.1). Such methods include all attempts to specify the principles of valid arguments, independent of their particular contents or contexts. In logic, those methods would include efforts to specify universally valid rules of inference. In mathematics, they would include methods of computation and proof. In empirical science, they include the specification of a unitary set of methodological principles believed to apply to different fields of research. The goal behind such generalization and formalization of methods is an increase in the certainty of the results obtained, an emphasis that leads in turn to the idealization of a particular mode of reasoning, namely, *deductive inference*. Given a set of premises known to be true and a set of valid rules of inference, one's conclusions follow with the certainty of logical necessity. Interpersonal argumentation is always to some extent context bound, and for this reason it lacks the rigor of deductive inference.

Another consequence of the move to formal methods is that reasoning loses much of its discursive character. Argumentation is always directed toward the beliefs and opinions held by particular persons, but formal methods involve the attempt to consider *possible* alternative viewpoints, even if they are not embraced by any real interlocutor. Such anticipation of purely hypothetical possibilities is a frequently mentioned characteristic of formal thinking (Inhelder & Piaget, 1955/1958). Thus, anyone mastering formal *methods* of argument will acquire in the process certain habits of formal *thinking*. In this view, formal thought develops when individuals apply formal methods to their own cognitive processes.

The implications of this thesis will become apparent in considering *everyday reasoning*. From Fig. 6.1, one might conclude that argumentation can be interi-

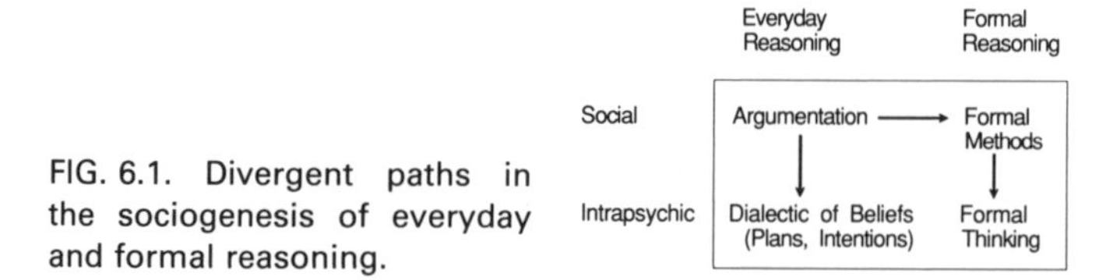

FIG. 6.1. Divergent paths in the sociogenesis of everyday and formal reasoning.

orized directly, not merely by way of formal methods. Such a conclusion suggests a further question regarding the kind of cognitive activity that results from such a direct interiorization of argumentation. One way of answering this question is to consider the definition of argumentation given earlier and what would change when this social, intellectual, and verbal activity is interiorized. The most obvious change is that it would no longer be social and verbal, but only intellectual. More substantively, it would be directed, not at resolving differences of opinion *between persons,* but at reconciling differences among *one's own* beliefs, plans, and intentions.

What remains of argumentation is something of the process by which such reconciliation occurs. Suppose one is confronted with a new datum that is inconsistent with some of one's existing beliefs. In order to resolve this inconsistency, one of the following alternatives must occur. Either one adopts the datum as a new belief and rejects or modifies those of one's existing beliefs that are inconsistent with it, or one rejects the new datum and retains one's existing beliefs more or less unchanged. Either way, the decision-making process would involve, not only the weighing of reasons for and against the beliefs in question, but also the possibility that some of those same reasons might be reevaluated. The point is not that human beings always make decisions in such a rational manner nor even that they *should* do so, but to describe how they proceed when they do evaluate beliefs in the light of reasons.

Such a process of internal deliberation is a dialectic of *beliefs* if it involves epistemic states, and it is a dialectic of *plans and intentions* if it is directed toward practical action. The process is dialectical not only because it begins with an opposition, but also because it preserves some of the functions of dialectical argumentation. In particular, it preserves the functions of *justification* and *refutation* to the extent that inconsistencies in one's beliefs (or plans and intentions) are resolved by considering the reasons that support and/or undermine the beliefs in question. The claim is not that other less rational factors do not also exist, only that, in addition to any such psychodynamic factors, human beings also possess the capacity for decision making based on a weighing of reasons. A further claim is that this capacity characterizes much of what one would like to include under the concept of everyday reasoning.

EVERYDAY REASONING AND DISCOURSE PROCESSES

One implication of the sociogenetic differentiation hypothesis is that everyday reasoning might retain more of its discursive origins than does formal reasoning. Some authors have argued that common fallacies of deductive reasoning might result from adherence to the norms of conversation like those described earlier (Politzer, 1986; Rumain, Connell, & Braine, 1983). For example, both children and adults are known frequently to commit the common fallacy of *affirming the consequent*. Given a conditional sentence such as, "If it rains, then the grass gets wet," and the additional information that the grass is in fact wet, subjects tend to draw the fallacious conclusion that it must have rained (e.g., Byrnes & Overton, 1988). This inference is a fallacy, because wet grass may have resulted from other possible causes. The common tendency even of adult subjects to commit such logical fallacies is one of the core phenomena that, according to Galotti (1989), ought to be explicable in any theory of everyday reasoning.

One explanation for such findings is that subjects interpret the information provided to them in problems of formal reasoning in terms of the norms of conversation mentioned earlier. For example, they might assume, consistent with the maxims of quantity and relation, that if other relevant causes of the grass getting wet existed, the speaker would have mentioned them. Under the assumption that rain is the only relevant cause of the lawn getting wet, the conclusion that it must have rained if the grass is wet becomes quite reasonable. Evidence for such an explanation of fallacious reasoning is provided by Rumain et al. (1983): Children tended to commit the fallacies of affirming the consequent and denying the antecedent when they were presented basic premise information, but not when they were presented additional information that countermanded the assumption that all the relevant information had been given. Their interpretation of these findings was that logical fallacies result from "conversational comprehension" based on Grice's maxims as described previously. According to Politzer (1986), most of the common errors of formal reasoning can be explained in terms of the conflict between the syntactic rules of logic and pragmatic rules of discourse such as Grice's maxims.

That thesis has important implications for developmental and cross-cultural research. Developmentally, it implies that children who have mastered the pragmatic rules of language use may have to unlearn some of these rules when it comes to formal reasoning. More precisely, they may have to learn that particular contexts exist in which those rules do not apply. According to Scribner (1977), formal reasoning requires mastery of a particular "genre" of discourse, and formal schooling may be a central means for acquiring a mastery of the "logical genre."

The potential conflict between the rules of language and of logic might also help to explain why the Central Asian peasants studied by Luria (1976) in the 1930s failed to reason syllogistically. For example, when told that there were no camels in Germany and that Berlin was in Germany, they refused to draw the seemingly obvious conclusion that there were no camels in Berlin. Instead, they tended to express uncertainty about whether there were camels in Berlin in the absence of relevant experience. Luria took this refusal to reason syllogistically as the mark of a practical mentality. Another interpretation is that his subjects were following Grice's maxim of quality, which prohibits one from making false or unfounded statements. Lacking a knowledge of the "logical genre," they focused on the empirical truth of their statements.

However, the relation between formal logic and discourse processes can be interpreted in different ways. One hypothesis is that the structure of intrapsychic inference can indeed be modeled on the theorems of formal logic, but that conversational comprehension processes can interfere with this underlying structure and prevent it from being expressed in performance (Braine & Rumain, 1983; Rumain et al., 1983). This view implies that the psychological processes involved in intrapsychic inference are distinct from those involved in conversational comprehension. But if some forms of intrapsychic inference develop directly from interpersonal argumentation, then the former might retain some of the characteristics of the latter. In other words, everyday reasoning might display some structural properties deriving from conversation and differing in important ways from the inferential structure of formal reasoning.

For example, argumentation theorists have pointed out that real arguments are based on forms of justification that are weaker than deductive inference (Rescher, 1977). One of the most common moves in real argumentation is an appeal to what might be called *presumptive evidence:* justifying a conclusion by appealing to what is known to be normally or typically the case, under the assumption that one is not dealing with an atypical instance. Take the statement, "Jennifer can read and write." This statement might be justified by the proposition that the person in question is a high school graduate, the presumption being that high school graduates typically have acquired basic literacy skills. Obviously, the presumption is not that *all* high school graduates have acquired such skills. The presumption is only that basic literacy is a *normal* or *typical* outcome of a high school education. So one may be justified in concluding that a person can read and write, as long as one can assume that the individual in question is a typical high school graduate in this respect. If, instead, evidence is produced which countermands this assumption, then the presumptive inference must be withdrawn.

NONMONOTONIC REASONING

The implication of the foregoing argument is that, if everyday reasoning develops in part from argumentative discourse, then it is likely to involve some of the same kinds of presumptive inferences found in argumentation. Reasoning according to such principles might differ in substantial ways from formal reasoning. The characteristics of assumption-based inferences of this kind have been studied in cognitive science under the rubric of *nonmonotonic reasoning* (Ginsburg, 1987; Reiter, 1987). The defining characteristic of such reasoning is that the number of valid conclusions that can be drawn does not increase monotonically with the addition of new information. Instead, new information might result in the retraction of previous conclusions based on assumptions now found to be false.

The main goal of theories of nonmonotonic reasoning has been to explicate certain characteristics of everyday, "commonsense" reasoning, particularly the ability to make inferences, not only from the information explicitly available, but also from the *absence* of certain kinds of information. Several approaches to this problem have been developed, including *closed world databases* (reasoning based on the assumption that all the relevant information is included in an existing database), *circumscription* (reasoning based on an assumption that exemplars of a given class are typical exemplars unless they are explicitly known to be atypical), and *default logic* (reasoning that allows conclusions to be drawn from incomplete information as long as those conclusions are consistent with the database). For present purposes, it is unnecessary to review each of these approaches in detail (see Ginsburg, 1987; Reiter, 1987). What they all have in common is the attempt to explain reasoning based on assumptions made *in the absence of information to the contrary.* For example, given the information that a certain person is a high school graduate and the absence of any information to the effect that the person in question is not a typical high school graduate, then one is justified in concluding (at least provisionally) that the individual has acquired basic literacy skills.

Reasoning based on assumptions of this kind lacks the certainty and necessity of deductive inference, but it compensates for this lack of rigor through its economy of representation. The fact that all the relevant information does not have to be represented explicitly gives default reasoning a certain practical advantage over deductive inference for reasoners with limited processing resources. Moreover, assumption-based reasoning has appeared to some theorists to be admirably suited for modeling the process of belief revision. For example, in Doyle's (1979) *truth maintenance system,* beliefs are classified as "in" or "out," depending on whether they can be justified with reference to the individual's existing beliefs. Assumptions are allowed in the form of beliefs that are accepted in the absence of other beliefs that contradict them. When new information is added to the system, some of those assumptions might have to be retracted, along with any other beliefs the acceptance of which depended on those assumptions.

As Doyle (1979, pp. 256–257) himself pointed out, this system of belief revision can be likened to dialectical argumentation in which the opposing sides are represented, not by differences of opinion between persons, but by opposing modules of beliefs. The argument advanced in this chapter is that this likeness is no mere analogy. If the kind of reasoning involved in belief revision develops from interpersonal argumentation, then one might well expect similar forms of inference in each case. Just as expressed opinions are subject to reasoned justification or refutation in argumentation, so are individual beliefs subject to being supported or undermined in the reasoned revision of belief. And just as argumentation often hinges on presumptive evidence, so does belief revision, according to Doyle, revolve about assumptions made in the absence of complete information.

Indeed, a deep kinship exists between nonmonotonic assumption-based reasoning and implicit conventions like Grice's maxims of conversation that are involved in interpersonal communication. The connection between default logic and such conventions has been pointed out by Reiter (1987, p. 180). For example, listeners following Grice's maxims of quantity and relation (to provide an appropriate amount of information that is relevant to the topic) would assume, in the absence of information to the contrary, that nothing essential for the speaker's meaning was left unsaid. The default in this case is the rule, "If something is not mentioned, then it doesn't have to be considered." In contrast, formal reasoning often requires one to consider an exhaustive space of possibilities, including those not explicitly mentioned. As argued previously, one reason why subjects might commit the fallacy of affirming the consequent when provided with conditional statements such as, "If it rains, then the grass gets wet," is because they make the default assumption that all the relevant causal information has been provided. In contrast, the formal-logical interpretation of this statement requires one to consider possible causes of wetness that are not stated explicitly.

The overall point is that the "logic of conversation" might not interfere with the "natural logic of reasoning," as conceived by Rumain et al. (1983). Instead, the logics of conversation and of everyday reasoning might have a common structure, based on the developmental continuity between them. In this context, theories of nonmonotonic reasoning can be regarded as attempts to elucidate that common structure.

REASONING AS BELIEF REVISION

A major problem with current theories of nonmonotonic reasoning is that they have been formulated for the most part in isolation from the psychology of reasoning. Most of these theories have been attempts to formalize intuitions of the theorists about the character of everyday reasoning. One might be justifiably concerned that important aspects of everyday reasoning could have been overlooked in the very attempt to formalize it. For example, Doyle's truth maintenance system was designed to keep a complete record of all the ways in which a particular belief has been justified in the past. The intent behind this convention was to provide the system with some consistency in its justifications over time, but one may reasonably doubt that human reasoners keep such a complete record of their past justifications.

The consequences of such limitations in attention and memory for reasoning have been explored by Harmon (1986). In place of the attempt to formalize processes involved in everyday reasoning, Harmon proposed a theory of reasoning as belief revision based in part on the following principles:

1. *Positive undermining:* One should reject a belief when one has some positive reason to do so.
2. *Conservatism:* One should retain a belief in the absence of any specific reason to reject it.
3. *Clutter avoidance:* One should not clutter one's mind with trivialities.

Taken together, these principles imply that reasoners should attempt to maximize the coherence of their beliefs with a minimum of *changes* and that they should do so *without keeping track* of the reasons involved in past revisions of belief. Keeping track would lead to an accumulation of potentially trivial information, which would be dysfunctional for reasoners with finite processing resources.

In support for the claim that human beings actually follow such principles of belief revision, Harmon (1986, pp. 35–40) cited the phenomenon known in the social psychology literature as "belief perseverance after evidential discrediting" (Ross & Anderson, 1982). In a typical experiment, subjects are given false feedback about their social judgments, then informed that this feedback in fact was false. Despite this explicit debriefing, they are nevertheless influenced in rating their own performance by the feedback they now know to be false! Such findings are generally interpreted as resulting from irrational "biases" or "heuristics" in the seeking, assimilation, or recall of information. For Harmon, the processes involved are "irrational" only when judged by the standards of formal logic and appear more reasonable in the light of his principles of belief revision.

Harmon's view of reasoning as belief revision is primarily *internalist* in orientation. His main interest was in the principles governing the processes occurring within individuals when they are engaged in revising their own beliefs. An alternative, *externalist* view is represented by Mackenzie (1989). Against Harmon, he argued that reasoning is primarily a social phenomenon and that the processes occurring within a single individual in belief revision derive from the social case. For example, the process of maintaining coherence among one's own beliefs has its original counterpart in the avoidance of inconsistency in linguistic interaction. Thus, reasoning is an inferential process by which the partners to a dialogue monitor their interaction and determine whether or not their current utterances are consistent with the commitments expressed in previous utterances. According to Mackenzie's "externalist" theory of reasoning, Harmon's principles of belief revision are just representations to oneself of certain principles governing linguistic interaction.

Mackenzie's view is consistent with the approach taken in this chapter, except for the concept of sociogenesis described herein as the developmental link between intrapsychic inference and interpersonal interaction. In conformity with the arguments summarized earlier, the externalist view implies (a) that reasoning is both a social, interactive process as well as an intellectual activity, and (b) that reasoning as a social phenomenon is in some sense primary and reasoning as intrapsychic inference secondary. The concept of sociogenesis provides a developmental explication of the latter proposition.

IMPLICATIONS

The approach to everyday reasoning described in the preceding sections has important implications for explaining the research findings that according to Galotti (1989) comprise the core phenomena to be addressed in any general theory of reasoning. Perhaps the most obvious implication has to do with the interpretation of the kinds of logical errors that are common in everyday reasoning. The hypothesis advanced in this chapter has been that intrapsychic inference develops from interpersonal argumentation. If this thesis is correct, then one might expect (a) that the inferential structure of everyday reasoning would be influenced by the structure of argumentation (and more generally, by the pragmatic rules of language use exemplified in Grice's maxims), and (b) that this structure might differ in substantial ways from that of formal reasoning.

Evidence implicating the logic of conversation in common errors in reasoning was provided by Rumain et al. (1983). In their study, children tended to commit the fallacies of affirming the consequent or denying the antecedent when they were presented single if–then statements (e.g., "If there is a duck in the box, then there is a peach in the box"), but not when they were presented with multiple if–then statements, two of which shared the same consequent (e.g., "If there is a dog in the box, then there is an orange in the box; if there is a tiger in the box, then there is an orange in the box"). These results were explained by assuming that "discourse comprehension processes" based on Grice's maxims cause children to commit the fallacies in the first condition but not the second. In the single-statement condition, children assume that if anything else went with the peach in the box, the speaker would have mentioned it. So they conclude that, if there is *no* duck in the box, then there is also no peach. In the complex-statement condition, the speaker does mention another antecedent for the orange in the box, so they are not tempted to conclude that there is no orange if there is no dog.

Rumain et al. viewed the "logic of conversation" upon which comprehension is based as distinct from the "natural logic" of human reasoning, the latter being understood in terms of inference schemas derived from among the theorems of formal logic. The alternate view stated in this chapter is that the structure of everyday reasoning more closely resembles the pragmatic structure of conversation than that of formal logic and that the structural properties of everyday reasoning and conversational discourse can be described in terms of assumption-based nonmonotonic reasoning. Although the nonmonotonic structure of everyday reasoning is less rigorous than that of formal reasoning, it can be adaptive for reasoners with limited processing resources because it minimizes the number of possibilities explicitly considered in making inferences. Despite the defeasibility of assumption-based reasoning, it can result in valid inferences to the extent that the assumptions upon which it is based are sound.

In this view, the development of formal reasoning has at least two necessary conditions. The *ontogenetic* condition is that children develop whatever processing resources are necessary for considering exhaustive sets of hypothetical possibilities (see Piaget, 1981/1987, or Chapman, 1988, pp. 316–322, on the development of children's understanding of possibilities). The *sociocultural* condition is that children (or adults) become acculturated into formal modes of discourse in which the truth values of statements based on hypothetical possibilities are considered. As described by Scribner (1977), formal schooling is the major means by which this "logical genre" of discourse is acquired. Formal reasoning develops only when both of those conditions are fulfilled, and that is why the reasoning of children who lack the ontogenetic prerequisites for formal reasoning cannot be equated with that of adults who lack the sociocultural prerequisites.

Another central idea expressed in this chapter was that everyday reasoning is not limited to problem solving, but is also likely to be involved in the revision of belief (Doyle, 1979; Harmon, 1986). In reasoning as problem solving, inferences are drawn from the existing knowledge base in order to solve particular problems. In reasoning as belief revision, the knowledge base itself is modified. As described by Doyle (1979) and Harmon (1986), belief revision is a process governed by principles of limited coherence and consistency. The process is limited, because human reasoners cannot survey the entire field of beliefs at one time in order to check for coherence among the entire set. Instead, consistency checks are limited to those subsets of the field which are small enough to be surveyable. Another limitation on this process is the previously described minimization of considered possibilities inherent in nonmonotonic reasoning. Taken together, these two limitations constrain the development of formal reasoning in particular ways.

For example, consider the common tendency of children and lay adults to ignore evidence inconsistent with their currently held theories. Such a tendency could result from the attempt to maximize the coherence among their beliefs (including both theories and evidence) by ignoring those pieces of evidence which do not cohere with their currently held theories. In this case, theories and evidence occur simply as propositions to be believed or not according to the extent to which they cohere with other beliefs. In this context, evidence may be rejected when it does not cohere with a currently held theory, or a theory may be rejected when it does not cohere with the evidence. But given the need for explanation, a currently held theory is unlikely to be rejected until an alternative explanation is available. The result is that evidence is often ignored or denied when it does not cohere with the current theory and no alternative is immediately at hand.

As Kuhn (1989) has argued, scientific thinking develops when subjects begin to differentiate theories from evidence and to view their currently held theories as only one among a set of *possible* theories, each of which coheres with a different pattern of data. One consequence of such a differentiation is that theories and evidence begin to be treated differently. Ideally, coherence is now obtained by rejecting theories that do not cohere with the data and not by rejecting data that do not cohere with currently held theories. This new asymmetry between theories and data becomes possible only when alternative theories are viewed as being available as a matter of course—that is, when a given theory is always viewed as one among a set of alternative possible theories.

The hypothesis that everyday and formal reasoning have some distinctively different structural features might help to explain why they have been found to be uncorrelated, but it also recalls the dilemma mentioned at the beginning of the chapter. The fact that everyday thinking often does not conform to the standards of formal logic is sometimes interpreted as implying that it is fundamentally irrational. That conclusion seems inescapable as long as formal logic is the standard for rationality. However, the argument implicit in this chapter has been that rationality is a considerably broader concept than logic. It includes all forms of decision making based on a deliberative weighing of reasons. Everyday thinking sometimes may be *illogical* in the sense that it is based on principles that differ from those of formal logic (see Politzer, 1986), but it is not *irrational* on that account. No doubt the certainty and necessity of formal reasoning are preferable to uncertainty, other things being equal. But in everyday life the *ceteris paribus* clause frequently cannot be satisfied, and one frequently must draw inferences from incomplete information. In such circumstances, one is likely to use forms of inference that are less rigorous (but perhaps more practicable) than formal logic. Theories of nonmonotonic reasoning can be regarded as attempts to understand the structural principles of such everyday forms of reasoning.

This view on the relation between logic and rationality is relevant for recent discussions on the rationality of science itself (e.g., Newton-Smith, 1981). On one side, philosophers from Bacon to Popper, Lakatos, and Lauden have endeavored to describe the *logic* of scientific methods. On the other side, authors who have questioned the logical foundations of science have been accused of "irrationalism." Another interpretation is that science cannot be reduced completely to principles of formal logic. It always retains a measure of everyday reasoning, but it does not lose its rationality on that account. Its rationality consists, not in any conformity to abstract logical principles, but in the attempt to settle differences of opinion through a weighing of evidence, especially (but not exclusively) that obtained through observation. In the scientific forum this process is one of argumentation through publications, and in the minds of individual scientists it is a dialectical process of belief revision.

ACKNOWLEDGMENT

Preparation of this chapter was supported in part by Operating Grant # OG0037334 from the Natural Sciences and Engineering Research Council of Canada.

REFERENCES

Barilli, R. (1989). *Rhetoric.* Minneapolis: University of Minnesota Press.

Berkowitz, M., Oser, G., & Althof, W. (1987). The development of sociomoral discourse. In W. K. Kurtines & J. L. Gewirtz (Eds.), *Moral development through social interaction* (pp. 322–352). New York: Wiley.

Braine, M. D. S. (1978). On the relation between the natural logic of reasoning and standard logic. *Psychological Review, 85,* 1–21.

Braine, M. D. S., & Rumain, B. (1983). Logical reasoning. In J. Flavell & E. Markman (Eds.), P. H. Mussen (Series Ed.), *Handbook of child psychology. Vol. 3: Cognitive development* (pp. 263–340). New York: Wiley.

Byrnes, J. P., & Overton, W. F. (1988). Reasoning about logical connectives: A developmental analysis. *Journal of Experimental Child Psychology, 46,* 194–218.

Ceci, S. J., & Liker, J. K. (1986). Academic and nonacademic intelligence: An experimental separation. In R. J. Sternberg & R. K. Wagner (Eds.), *Practical intelligence* (pp. 119–142). Cambridge, UK: Cambridge University Press.

Chapman, M. (1988). *Constructive evolution: Origins and development of Piaget's thought.* Cambridge, UK: Cambridge University Press.

Doyle, J. (1979). A truth maintenance system. *Artificial Intelligence, 12,* 231–272.

Furth, H. (1981). *Piaget and knowledge* (2nd ed.). Chicago: University of Chicago Press.

Galotti, K. (1989). Approaches to studying formal and everyday reasoning. *Psychological Bulletin, 105,* 331–351.

Ginsburg, M. L. (1987). Introduction. In M. L. Ginsburg (Ed.), *Readings in nonmonotonic reasoning* (pp. 1–23). Los Altos, CA: Morgan Kaufmann.

Glass, A. L., Holyoak, K. J., & Santa, J. L. (1979). *Cognition.* Reading, MA: Addison-Wesley.

Grice, H. P. (1975). Logic and conversation. In P. Cole & J. L. Morgan (Eds.), *Syntax and semantics. Vol. 3: Speech acts* (pp. 41–58). New York: Academic Press.

Harmon, G. (1986). *Change in view.* Cambridge, MA: The MIT Press.

Hutchinson, R. M. (Ed.). (1952). *Great books of the Western world. Vol. 8: Aristotle, I.* Chicago: Encyclopedia Britannica.

Inhelder, B., & Piaget, J. (1958). *The growth of logical thinking from childhood to adolescence.* New York: Basic Books. (Original work published 1955)

Johnson-Laird, P. (1983). *Mental models,* Cambridge, MA: Harvard University Press.

Kuhn, D. (1989). Children and adults as intuitive scientists. *Psychological Review, 96,* 674–689.

Laboratory of Comparative Human Cognition. (1983). Culture and cognitive development. In W. Kessen (Ed.), P. H. Mussen (Series Ed.), *Handbook of child psychology. Vol. 1: History, theory and methods* (pp. 295–356). New York: Wiley.

Leadbeater, B. J. (1988). Relational processes in adolescent and adult dialogues: Assessing the intersubjective context of conversation. *Human Development, 31,* 313–326.

Luria, A. (1976). *Cognitive development.* Cambridge, MA: Harvard University Press.

Mackenzie, J. (1989). Reasoning and logic. *Synthese, 79,* 99–117.

Matlin, M. (1983). *Cognition.* New York: Holt, Rinehart & Winston.

Newell, A. (1981). Reasoning, problem-solving, and decision processes: The problem space as a

fundamental category. In R. Nickerson (Ed.), *Attention and performance* (pp. 693–718). Hillsdale, NJ: Lawrence Erlbaum Associates.
Newton-Smith, W. H. (1981). *The rationality of science*. London: Routledge & Kegan Paul.
Osherson, D. N. (1975). Logic and models of thinking. In R. Falmagne (Ed.), *Reasoning: Representation and process in children and adults* (pp. 81–92). Hillsdale, NJ: Lawrence Erlbaum Associates.
Piaget, J. (1928). *Judgment and reasoning in the child.* London: Routledge & Kegan Paul. (Original work published 1924)
Piaget, J. (1987). *Possibility and necessity. Vol. 1: The role of possibility in cognitive development.* Minneapolis: The University of Minnesota Press. (Original work published 1981)
Politzer, G. (1986). Laws of language use and formal logic. *Journal of Psycholinguistic Research, 15,* 47–92.
Reiter, R. (1987). Nonmonotonic reasoning. *Annual Reviews of Computer Science, 2,* 147–186.
Rips, L. J. (1983). Cognitive processes in propositional reasoning. *Psychological Review, 90,* 38–71.
Rescher, N. (1977). *Dialectics*. Albany, NY: State University of New York Press.
Rosch, E. (1987). Wittgenstein and categorization research in cognitive psychology. In M. Chapman & R. A. Dixon (Eds.), *Meaning and the growth of understanding: Wittgenstein's significance for developmental psychology* (pp. 151–166). Berlin: Springer-Verlag.
Ross, L., & Anderson, C. A. (1982). Shortcomings in the attribution process: On the origins and maintenance of erroneous social judgments. In K. Kahneman, D. Slovic, & A. Tversky (Eds.), *Judgment under uncertainty* (pp. 129–152). Cambridge, UK: Cambridge University Press.
Rumain, B., Connell, J., & Braine, M. D. S. (1983). Conversational comprehension processes are responsible for reasoning fallacies in children as well as adults: *If* is not the biconditional. *Developmental Psychology, 19,* 471–481.
Scribner, S. (1977). Modes of thinking and ways of speaking: Culture and logic reconsidered. In P. N. Johnson-Laird & P. C. Wason (Eds.), *Thinking* (pp. 483–500). Cambridge, UK: Cambridge University Press.
Scribner, S. (1986). Thinking in action: Some characteristics of practical thought. In R. J. Sternberg & R. Wagner (Eds.), *Practical intelligence* (pp. 13–30). Cambridge, UK: Cambridge University Press.
Sternberg, R. J. (1983). Components of human intelligence. *Cognition, 15,* 1–41.
Van Eemeren, F. H., & Grootendorst, R. (1982). The speech acts of arguing and convincing in externalized discussions. *Journal of Pragmatics, 6,* 1–24.
Van Eemeren, F. H., Grootendorst, R., & Kruiger, T. (1984). *The study of argumentation*. New York: Irvington.
Vygotsky, L. (1978). *Mind in society.* Cambridge, MA: Harvard University Press.
Wagner, R. K., & Sternberg, R. J. (1986). Tacit knowledge and intelligence in the everyday world. In R. J. Sternberg & R. K. Wagner (Eds.), *Practical intelligence* (pp. 51–83). Cambridge, England: Cambridge University Press.
Wertsch, J. V., & Stone, C. A. (1985). The concept of internalization in Vygotsky's account of the genesis of higher mental functions. In J. V. Wertsch (Ed.), *Culture, communication, and cognition* (pp. 162–179). Cambridge, UK: Cambridge University Press.
Wittgenstein, L. (1958). *Philosophical investigations*. New York: Macmillan.

IV EVERYDAY COGNITION ACROSS THE LIFE SPAN

7 The Contextual Nature of Earliest Memories

Stephen J. Ceci
Cornell University

Helene Hembrooke
SUNY Binghamton

There was a time when the study of infant and young children's memory was the purview of a small group of us who were doing basic research on the mnemonic processes that were developing over the first few years of life. The mid to late 1970s were the heyday of such research in terms of the quantity of studies that were conducted and subsequently published. During those years, we learned so much about the development of strategies, knowledge, and insights that some of us deluded ourselves into believing that the memory development riddle had been solved. As one indication of this, the reader can examine any of the three main developmental journals during that period (*Child Development, Journal of Experimental Child Psychology,* and *Developmental Psychology*) to verify that the late 1970s was the apogee of memory development publications. If one searches titles of published abstracts for key words that have to do with memory development (e.g., rehearsal, organization, eidetic imagery, LTM, STM, decay, reminiscence, strategies, trace, semantic knowledge, capacity differences), it is clear that this type of research peaked around the late 1970s. If we compare the number of articles that were concerned with memory development during the late 1970s and early 1980s, the unmistakable trend is one of declining frequency of published memory development articles, starting in 1983, but probably reflecting a decline in the conduct of such studies that began around the late 1970s—since it usually takes several years between the initiation of a study and its fruition as a journal article (see Fig.7.1).

The cause for the decline in memory development research is not apparent in Fig. 7.1, and some might suspect that it is due to conscious editorial policies that shifted away from publishing basic memory papers, especially developmental ones, or perhaps to conscious decisions on the part of funding agencies not to

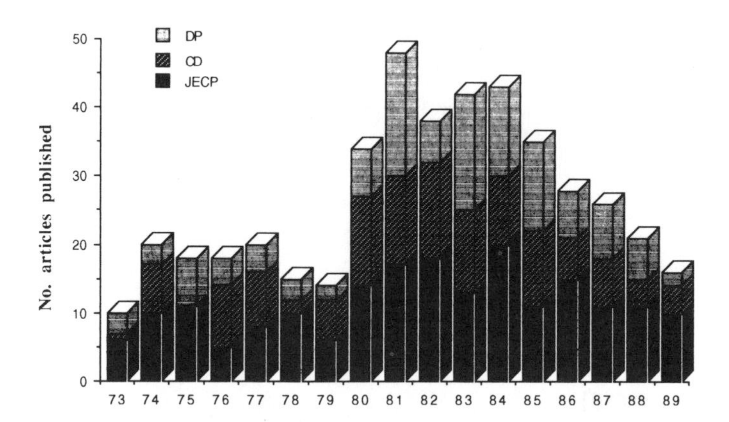

FIG. 7.1. JECP = Journal of Experimental Child Psychology; CD = Child Development; DP = Developmental Psychology. (Years of publication on abscissa.)

finance memory development research. But this explanation is almost certainly not the reason for the decline because there were no obvious editorial shifts away from publishing basic memory research during this period, nor were there any obvious program shifts at NSF and NIH that disproportionately jeopardized memory development vis-à-vis other types of developmental studies. Those of us who contributed to the flood of memory papers in the '70s and early '80s, however, can share with the reader a different explanation for the decline in memory development articles: Many of us ceased working on memory development problems out of a feeling that the really interesting questions had been answered—or so we thought.

As a traditional memory development research who abandoned this area in the early 1980s, the first author can share a personal experience that prompted his return to this area in the mid-1980s. Something unexpected happened around this time: Attorneys, child protective service workers, and judicial policy-makers began approaching memory development researchers for their advice about children's memories. The context that was most frequently asked about was that of sexual abuse. Children were going into courts in ever-increasing numbers to offer their uncorroborated recollections about who did what to them, and when and where it allegedly occurred.

It came as a surprise to some of us that although we prided ourselves on being child memory experts, we felt powerless to address this aspect of everyday cognition. Others have expressed this same feeling, including Peter Ornstein and Charles Brainerd, to name two distinguished contributors to the memory development area during the 1970s.

How is it that we memory researchers had convinced ourselves that we had solved the memory development riddle when in actuality we knew relatively little about the factors that affected the everyday types of remembering that brought young children into court? To answer this question we take a circuitous route, first, describing some recent trends in legal procedures that have resulted in a greater demand for expert testimony from child memory researchers, then providing a review of both data and theory about the earliest memories. We draw on some animal research because it may provide interesting new evidence that the human memory system is highly responsive to contextual factors, a point to which we return later. Finally, we conclude by suggesting various factors that may be involved in infantile amnesia.

RECENT TRENDS IN CHILD WITNESSES

In the decade of the 1980s, all 50 states had relaxed their standards for allowing children to testify about abuse. Currently, children can offer their recollections to courts without the requirement of corroboration by another witness or material evidence. As just one example of the trend that led states to drop corroboration rules that required the testimony of another witness to verify a child's story about abuse, the State of New York is revealing: In 1982, the last year before New York modified its corroboration rule to allow children to testify without corroboration, there were over 1100 reports of child abuse to the New York State Maltreatment and Abuse Hotline in Albany, of which only 180 could be "indicated." The designation *indicated* is used when a preliminary investigation reveals grounds for further investigation and possibly legal action, usually the availability of corroborating witnesses and/or medical evidence. Of the 180 *indicated* cases, only 11 were prosecuted in court, and of these, only 5 resulted in convictions (Abrams, 1984).

The majority of the 1100 cases never went to court because there was no corroboration of the child's claim and no medical evidence to support it. (However, the absence of litigation of the majority of the indicated cases may not have been due to a lack of corroborating testimony, but rather to a determination by social services personnel that litigation was not in the best interests of the child, and therefore the defendants were given the option of counseling, supervised visitation, plea bargains to noncriminal offenses, etc., as alternatives to litigation.) But no matter how one chooses to gainsay such data, there is no question that many cases of real abuse existed among the 1000 cases and were not pursued by law enforcement professionals due to the absence of a witness who could corroborate what the child told the law enforcement authorities.

In view of the vast number of claims of sexual abuse that began coming to light around 1980 (estimated by the National Center for Child Abuse and Neglect in 1986 to be 132,000 actual indicated cases per year—see APA amicus brief in Craig v. Maryland), and the inability of law enforcement officials to prosecute these cases effectively if the child's uncorroborated recollections were not allowed to stand, states abandoned their requirement that the child's testimony be corroborated. This decision opened the floodgates of young children into our nation's criminal courts, resulting in attorneys and policy-makers approaching memory development researchers for expert testimony about factors that influence the veracity and durability of children's earliest memories.

A perusal of the depositions and in-court testimony of some memory development experts reveals the difficulty they seem to have in generalizing from the basic research corpus on memory development to a particular case (Ceci & Bruck, in press). After all, what can an expert witness draw on from the decades of basic memory development research that would allow him or her to offer an opinion as to the mnemonic veracity of a 33-month-old who claims that the defendant touched her genitals while reading Little Red Riding Hood to her 55-month-old sister and her a year earlier, when she was 21-months-old and her sister was 43-months-old? And suppose, further, that the older sister has no recollection of ever seeing the defendant touch her sister while reading the story in question.

Even a brief comparison of this context with the typical basic research contexts of the 1970s drives home the multitude of differences, some of which may be quite important. For instance, earlier research is, according to Ornstein and his colleagues, insufficient for determining whether a 3-year-old's memory is superior for an event he or she participated in as opposed to merely observed (Ornstein, Gordon, & Larus, 1992). Furthermore, there exists very little basic memory development research that contrasts the accuracy of 3-year-olds recalling actions that occurred to their own bodies versus those they passively witnessed occurring to others (Goodman & Reed, 1986; King & Yuille, 1987). Yet, such factors as event participation and locus of the action now have been shown to be important in judging whether age differences will be found between 3- and 5-year-olds (Goodman, Rudy, Bottoms, & Aman, 1990). In some contexts, 3-year-olds' memory has been shown to be as accurate as older children's, and children's memory for actions has been shown to be superior to their memory for other events (King & Yuille, 1987).

So, those of us who "jumped ship" in the early 1980s out of a sense that the important questions had been answered, experienced a regression to reality upon discovering that we were unable to offer even the most rudimentary advice to jurors about children's remembering in those contexts that brought them to court. In summarizing the inadequacy of basic laboratory studies of memory development for generalizing to everyday instances of recollections, Ornstein, Larus, and Clubb (1992) commented:

> The type of memory that is involved in testimony situations requires recollection of personally experienced or witnessed events that are most often considered very meaningful, stressful, and possibly traumatic. In contrast, laboratory studies of children's memory often involve the use of word lists, pictures, toys, and stories as stimulus materials. Although many important general principles of memory can be

derived from these investigations, little information is available concerning children's long term recall of salient, personally-experienced events. (p. 9)

Memory in Context

Perhaps the one article of faith that has emerged from the attempt to understand memory in everyday contexts is that it can only be poorly approximated if we rely exclusively on the contextually sanitized laboratory settings in which we have traditionally studied children's memories. Everyday memory is more "restless" than the passive recounting of emotionally neutral events under contextually deprived conditions: it entails a seamless merger of social and cognitive forces (Ceci, 1990; Ceci, DeSimone, & Johnson, in press). For example, Ceci, Ross, and Toglia (1987) showed that persistent erroneous suggestions to children about what they allegedly observed had a more deleterious effect on the memories of 3-year-olds than on older children's memories, but this effect was exacerbated when the person making the false suggestions was perceived as a powerful authority figure (an adult) rather than as another child. Thus, social factors merged with cognitive ones to produce patterns of recall (see Fig. 7.2).

One of the most striking differences between the research most of us did in the '70s and what is being done today by those who refer to themselves as psycholegal researchers is the insertion of social factors into studies of memory. A case in point is the research of Gail Goodman. She was studying action memory in the late '70s and published an influential article in *Cognitive Psychology* on this topic that appeared to say all that we needed to know (Goodman, 1980). Today, Goodman still studies action memory; but she does this in the context of affect, motivation, and a variety of contextual factors. In short, she has seen the need to step outside traditional cognitive boundaries and assess children's memory for acts perpetrated against their bodies; she frequently has reported dramatic differences in recallability when children are assessed in such contexts vis-à-vis emotionally neutral ones. Many of us have now provided evidence that social factors influence not only a child's manner of solving a memory problem, but also her

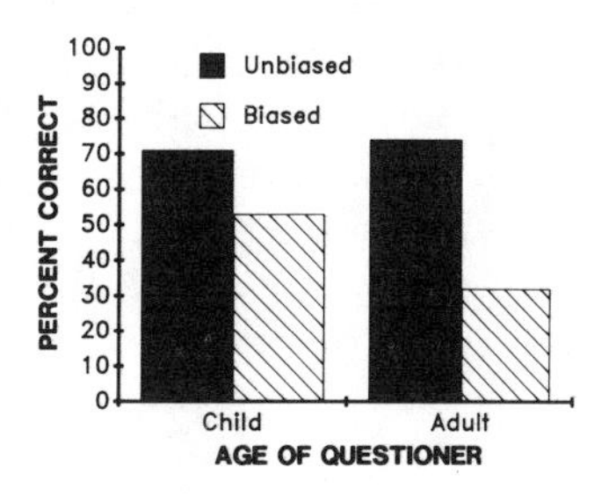

FIG. 7.2. Percentage of errors on a recognition memory test as a function of whether the person making the erroneous suggestions to a child was another child or an adult. ("Bias" refers to data when the interviewer made erroneous suggestions; "standard" refers to the control condition wherein no erroneous suggestions were made.)

perception of it (e.g., Ceci & Bronfenbrenner, 1985). For instance, some tasks may induce boys to employ different strategies from girls—despite the ability of both to use the same strategies—because of their differing perceptions of the task's social demands (i.e., whether it is a male or female sex-typed task). As we shall see later, there is now evidence from animal memory studies that contextual factors, including the social and emotional context in which acquisition occurs, can be extremely important in determining what is learned and how easily it will be forgotten.

The use of naturalistic contexts to study memory is less a rupture with tradition than a departure from the very recent ways that memory researchers have plied their trade. The seminal memory researcher who strove to understand processes in the absence of context was, of course, Hermann Ebbinghaus. The impression one has of Ebbinghaus is that he tried to expunge the memory task of all aspects of context, and hence ended up creating 2,300 nonsense syllables which he then proceeded to attempt to memorize himself. Yet, this is only one side of Ebbinghaus. There is another. He fully understood that memory was affected by social conditions and he wrote about this explicitly in the Preface to his book on memory (Ebbinghaus, 1885). Ebbinghaus acknowledged the importance of studying contextual determinants of memory after he finished plotting the basic parameters of short term and long term retention and savings. Unfortunately, he died before he ever did this.

Following Ebbinghaus, with the exception of researchers such as Bartlett (1932) and Istomina (1945, but not published until 1975), the habit until the 1970s was to pursue memory out of context, to study basic processes as they appear under highly controlled and unnatural conditions. Today, this practice is changing, and it is changing very rapidly (Ceci & Bruck, in press). An assumption being made in many quarters is that context not only serves to shape a child's perception of a memory problem, but also influences the solutions he or she deploys to solve it. Thus, "context not as an adjunct to cognition, but as a constituent of it" could be the motto of researchers who are revisiting old questions in new ways (Ceci, 1990). We have recently summarized this trend, as follows.

There is a "new look" in memory development research, and it is decidedly contextual. The crux of the current view, in fact, is that memory processes cannot be adequately understood or evaluated acontextually: To think about memory without considering the context that lead children to remember is akin to thinking about smiles independently on the faces on which they appear. Different contexts not only evoke different strategies to aid recall, but they also differentially shape an individual's perception of the recall task itself. Depending on the context in which remembering takes place, children may recall everything or nothing; their level of performance speaks as much to the power of context as to their native mnemonic capacity. (Ceci & Leichtman, p. 223, 1992)

EARLY MEMORIES AND AMNESIAS

The inability of older individuals to remember events that occurred before the age of 3 is a phenomenon referred to as "infantile amnesia." This void in early memory presents somewhat of a paradox to developmentalists because we often assume that early experiences influence later behavior and may even be associated with later forms of psychopathology such as depression. Thus, while we agree that the environment during early development is important, documenting children's ability to process and retain experiences from their environment has proven to be quite difficult.

Freud (1963) asserted that infantile amnesia was the result of repression. Repressed memories were at the root of neurosis and hysterical disorders, and the goal of therapy was to recover these memories. This process of recovering dormant memories involved analyzing the fragments of memories that are retrievable, but which "screen out" or protect the patient from the emotionally laden contents of the "forgotten" memories, usually assumed to be sexual or aggressive in nature.

As distinctly human as this may seem, infantile amnesia is not unique to humans. When trained to equivalent levels of learning, infant rats forget more rapidly than do adults. Theories to account for this phenomenon range from neurological deficiencies, to storage and retrieval factors, to encoding differences between adults and infants. A full review of this literature is beyond the scope of this chapter, and the interested reader is referred to reviews of the animal learning literature by Spear (1979a), the adult neurobiological literature by Squire and Cohen (1984), and the childhood neurobiological literature by Howe and Courage (in press). Here we focus on only those issues from the animal memory literature that seem especially pertinent to our main concern with the role of contextual factors in human memory development.

Before doing so, however, it is useful to make one point. The study of infantile amnesia entails not only a set of interesting theoretical questions that memory developmentalists have not resolved, but it also involves some thorny ethical issues, such as the use made of social science research by courts. In many cases across the United States and Canada the issue of infantile amnesia crops up. For example, in Moran versus the Commonwealth of Massachusetts, the defendants, William Moran and his wife were accused by their two grandchildren of molesting them while their parents were out of town looking for a house. The youngest grandchild was around 20-months-old when the alleged abuse occurred at her grandparents' home in Cambridge, MA. She recollected it for the first time when she was nearly 4-years-old, following a therapy session of her mother's, which she attended. The question at court was whether her "memory" of abuse was accurate; is it possible for a 4-year-old to recollect what happened to her when she was 20-months-old, or were the powerful adult suggestions via therapy instrumental in planting a false memory? One thing is very likely: Research on this question will find its way into courts, whether it is described by the researcher, complete with the usual caveats and qualifications about external validity, or by someone who has only heard about it from second-hand sources. Great care is needed in how researchers describe their results to minimize the chance that they will be misused by those involved in the adversarial process.

Studies of Mice and Men

Driven by the explanation that infantile amnesia results from a retrieval failure, researchers have documented that accelerated forgetting in young rats can be alleviated by exposure to a brief reminder of the training situation prior to memory testing. Campbell and Jaynes (1966) were the first to demonstrate this, showing that the representation of a CS or UCS during the retention interval could significantly reduce forgetting. As Spear and Parsons (1976) pointed out, this finding has been frequently replicated, using a variety of paradigms and retention intervals, with humans as well as animals. In the human research, it appears that 3- and 6-month-olds' memories are extremely dependent on the cues and contexts of training and testing. Cues that are similar, yet different, from those used during acquisition, will not serve as effective reminders for infants of either age. Similarly, changing the environmental context at the time of testing will also disrupt memory (Rovee-Collier, 1990).

The importance of the extrinsic aspects of context is only part of the story, though. Reactivation procedures have also been designed to reexpose subjects to the intrinsic cues present at the time of acquisition. In an early series of experiments, Haroutunian and Riccio (1979) exogenously administered substances that the body naturally releases when the organism is stressed or frightened. Injections of both epinepherine and acetylcholine as *reminders* 1 week prior to testing alleviated forgetting of a learned discriminative fear response in rats. However, neither substance was effective unless the animals were also reexposed to apparatus cues. In addition, aversive stimuli other than the UCS were effective reminders, suggesting that they too are successful in reinstating the emotional context.

The contribution of context to memory can also be assessed from a somewhat different angle. Nonassociative aspects of the training situation can also influence learning and memory. Several studies have reported dramatic improvements in retention in neonate and infant rats when tests were conducted in the presence of certain cues associated with the rat pup's nest. For example, Smith and Spear (1978) reported that the learning deficits typically seen in young rats on passive avoidance tasks and discriminative escape tasks were virtually nonexistent when home litter shavings were present under the apparatus. Rudy and Cheatle (1979) demonstrated retention of an odor aversion task after 6 days, for 2-day-old rat pups. This is remarkable in view of other research that has argued that the rat is incapable of even a 24-hour retention of simple tasks before it is at least 9 days old (cited in Spear, 1979b). What was different in the research of Rudy and Cheatle is that the pups were conditioned in the presence of littermates, making it an unusual context in the animal learning literature. This finding has since been replicated by Smith and Spear (1980). Recently, Lickliter and Hellewell (1992) have shown that the social context in the nest of young hatchlings is a powerful determinant of subsequent auditory recognition of maternal calls—more powerful perhaps than maturational schedule. These results, and many others like them, suggest that familiarity is an important contextual determinant of early memory, at least under some circumstances. (For additional demonstrations of the influence of familiarity on children's memory strategies, see Accredolo, 1979; Ceci & Bronfenbrenner, 1985.)

Thus far the role of contextual factors has been explored in terms of retrieval mechanisms. However, contextual factors may be influential even at the time of encoding. For example, it has been suggested that exaggerated forgetting in human infants may be due to encoding processes that result in memory representations that are qualitatively different from those of adults. Contrary to earlier claims, infants do not encode less than adults about a conditioning episode, but they actually appear to encode more. This may seem counterintuitive in view of their more rapid forgetting than adults. But the problem lies in what is encoded. It appears that although infants may encode more information than adults, they are not as proficient as adults in selecting the stimulus attributes that are associated with important consequences, that is, the UCS (Spear, Kucharski, & Miller, 1989).

Though it hardly seems adaptive to be relatively indiscriminant during encoding, infants actually appear to process their inordinate information load in an efficient manner. Research suggests that young animals have a built-in disposition to unitize or configure stimulus elements that are experienced in different modalities. This is evidenced by the finding that with mutlielement stimuli infants show less blocking, and more potentiation than do adults (Spear, Kraemer, Molina, & Smoller, 1988; Spear & Kucharski, 1984). This tendency is not unlike E. Gibson's view of unity in perception, and is similarly related to other phenomena such as cross-modal transfer, expectancy, and synesthesia. Canfield and Haith (1991) has shown that 4-month-olds can extract a pattern from an alternating sequences of faces that clearly indicates their ability to form expectancies. For instance, if the sequence of alternating faces is right, right-right-left, then infants who have been exposed to this sequence will begin looking to the left side even before the face appears there, suggesting they "count" or unitize their environment in an extremely effective manner.

What is most important about the early cognitive development research for the present purposes is how this organization takes place. There is evidence that encoding in infants may proceed amodally, that is, based on amodal characteristics of stimuli such as familiarity, affect, and intensity rather than modality-related dimensions like color (Lewkowicz & Turkowitz, 1980; Spear & Molina, 1987).

Infants then, appear to encode more than the specific attributes of the nominal stimulus. Perhaps in an effort to cope with all this information, the infant integrates or unitizes the stimulus information by extracting larger patterns, and forming expectancies. The context in which this information may be organized might be qualitatively different from the code used by adults to construct their representations, though at present this is little more than a hypothesis awaiting empirical support.

The foregoing research on infantile amnesia suggests that the memory difficulties of both human and animal infants are strongly influenced by contextual factors, both intrinsic (e.g., the infant's affective state during acquisition) and extrinsic (e.g., features of the learning environment). This, and the research reviewed next, demonstrates that context can not be treated as a separate issue or a potential influence only in rare circumstances. Contextual influences on early remembering and forgetting are the rule, not the exception.

Very Long-Term Memory in Young Children

Numerous experiments have demonstrated long-term memory in infancy, although the retention periods explored have seldom exceeded a few days. In a review of the literature prior to the 1980s, Werner and Perlmutter (1979) noted that the retention intervals commonly explored have been limited to several minutes. However, somewhat longer intervals have been tested within the visual recognition memory framework. Fagan's research, for example, indicates that 5-month-olds' recognition of faces lasts up to 14 days, while for 7-month-olds it can last up to 2 days for abstract stimuli (Fagan, 1971, 1979). In a similar vein, Topinka and Steinberg (1978) indicated a 2-week-long retention period for abstract patterns in 7-month-olds, and Strauss and Cohen (1980) reported 24-hour retention of such patterns at 5 months. Howe and Courage (in press) have provided a review of the most recent research on very long term memory in young children. In the past few years there have been a number of studies of infants' and young children's remembering that have employed delay intervals of months and even years; without exception, these studies have yielded evidence for some types of memory, though the exact mechanisms involved in these studies are still unknown (Bauer & Mandler, 1990; Cutts & Ceci, 1988; McDonough & Mandler, 1990; Meltzoff, 1988a; Myers, Clifton, & Clarkson, 1987; Perris, Myers, & Clifton, 1990).

Rovee-Collier and her colleagues have used a conjugate reinforcement paradigm to study retrieval in young infants. In this procedure, infants learn a kicking response that produces movement in a mobile suspended above them. Memory is evidenced when the infant emits the kicking response after varying time intervals between acquisition and testing. Three-month-olds exhibit retention for this contingency for 6–8 days. Six-month-olds are able to remember it for 21 days (Rovee-Collier & Hayne, 1987).

Rovee-Collier's work has extended beyond the identification of these basic forgetting functions to explore the effects of various reactivation treatments on infant memory. Findings from her work indicate that when given a brief reminder 24 hours prior to test, infants exhibit near perfect retention after intervals in which the response is ordinarily forgotten (Hill, Brovosky, & Rovee-Collier, 1988; Rovee-Collier, 1984). In addition, as indicated earlier, the memories of infants this age are highly context-specific. Rovee-Collier's work shows that changing either the mobile or the test context will disrupt memory after a delay of only 1 day (Rovee-Collier, 1990). After longer delays, however, infants will transfer their learning to novel mobiles or contexts. This suggests that the general features of the stimulus and context are retained for longer periods of time than are the specific attributes. This same memory for general features was demonstrated by Nelson and Ross (1980) in their study of infants' and young children's event memory. Young children in their study exhibited good memory for general information, and reported significantly fewer specific instances of an event. The effects of reactivating memory has also been demonstrated with toddlers' long-term memory for a novel event (Hudson, 1990). Eighteen-month-olds exhibited greater recall after 8 weeks if their received a reminder of the event/activities during the retention interval.

Although a few examples have been given, reproductive processes, including recall, have been less frequently studied in prelinguistic children than has the much simpler forms of retention. Tests of reproductive processes tap a more active sense of the memory system, as they require some manner of re-living (or reproducing) a prior stimulus or event on the part of the subject. These processes clearly imply reference to a cognitive representation of the past that is not presently available, unlike the conceptually simpler recognition tasks (Mandler, 1984). Several paradigms for looking at such processes in infancy have been developed. One example of this is found in experiments focusing on infants' abilities to process and store event sequences or routines. Such work has revealed that infants between the ages of 5 and 18 months have the capacity to become familiar with routines and use them for prediction to varying extents, clearly implicating long-term memory processing (Kessen & Nelson, 1978; Smith & Hull, 1984). Related work has indexed the degree of surprise demonstrated by infants in various situations, using this as a measure of remembering. LeCompte and Gratch (1972), for example, reported a gradual increase in surprise in infants between the ages of 9 and 18 months upon encountering novel objects in places in which familiar objects had been hidden. In such experiments, however, surprise is difficult to measure, and response to novelty is easily confounded with the violation of a previously established expectation.

Another manner of studying reproductive memory has involved the use of deferred imitation. From a cognitive perspective, imitation requires the individual to produce a behavior experienced in the past, a feat that is arguably more difficult than merely recognizing a previously viewed stimulus or behavior. Recent work by Meltzoff on this topic has revealed several findings of interest. Meltzoff (1988a) has demonstrated that 9-month-olds can imitate simple actions incorporating novel objects when they are demonstrated by a researcher. The same set of studies has shown that infants can exhibit similar imitative behavior when a 24-hour retention interval is imposed between the researcher's modeling and the deferred test of infant modeling behavior. In this way, significant recall of the researcher's behavior was evidenced by the infants over a day-long period. Even more strikingly, Meltzoff (1988b) studied imitation in 14-month-olds with a retention interval of 7 days. In this study, six actions, each perpetrated upon a different object, were shown to each infant sequentially. The infants were not permitted to interact with any of the objects at the time of this demonstration, or during the retention interval that followed. One of the six actions was an unusual behavior that had no probability of occurring spontaneously during play. After the week-long delay, infants were permitted to interact with the objects on their own. The findings show that infants who had witnessed the earlier modeling spontaneously produced significantly more of the modeled acts than infants in a control group who had not witnessed the modeling. In addition, the unusual novel behavior was imitated by a large proportion (67%) of the infants in the experimental condition and, as expected, was not exhibited by any of the control subjects. As Meltzoff (1988b) pointed out, the week-long delay interval used in this study effectively eliminates a nonrepresentational explanation for the infant behavior, bolstering the notion of significant reproductive capacities in infants just over 1-year-of-age.

Also using the technique of modeling, McDonough and Mandler (1990) focused on the ability of 2-year-olds to remember actions that they observed or performed when they were only 11-months-of-age. Subjects from an experiment involving deferred imitation were brought back into the laboratory one year later. In the original experiment, subjects had viewed the modeling of either familiar or novel actions, using props representing familiar objects. They were given the opportunity to imitate these actions immediately and/or after a 24-hour delay. In the session taking place the following year, the same infants were allowed to manipulate each object prior to any subsequent modeling. Comparison with a control group, consisting of subjects who did not participate in the original experiment, indicated no overall between-group differences in the number of target actions performed by the infants. However, for one particular familiar action, feeding a stuffed animal with an abstract bottle, there was a significant effect. Subjects in the group that had originally witnessed this action produced it more often than subjects in the group that had not. The researchers concluded that this action, which was observed only twice and performed twice at most, could be recalled by 23-month-olds after a 1 year delay interval. Given the length of this interval, it is truly amazing that memory proved so durable. This finding adds an important piece of evidence to the infant long-term memory literature.

In line with these findings regarding the potential of infant reproductive memory, Cutts and Ceci (1988) reported that children as young as 8-months-of-age were able to reproduce a learned behavior 4 months after it was originally taught to them. In their study, infants were originally exposed to a puppet that had a mitten which, if removed, would reveal a treat. They were initially shown that the mitten could be pulled off the puppet to retrieve the treat, and they grasped the contingency between pulling it off and eating the treat almost immediately. After 3 weeks of extensive practice at removing the appropriate mitten, the puppet was taken away for 4 months. When it was presented to the infants again, only a disappointing 5% of the children appeared to remember that one of the mittens could be removed to recover a treat. (No control group child ever removed the appropriate mitten, or touched it.) Most of the children reacted indifferently to the puppet, preferring to play with other toys in the room. Nonetheless, when they did interact with the puppet, these same children engaged in reliably more touching of the appropriate mitten. In fact, they almost never touched the mitten that had not been associated with the treat 4 months earlier. They touched the appropriate mitten so much that it fell off within 5 minutes of playing in almost half of the cases. However, in those cases in which the mitten fell off, the children rarely looked for the treat, and did not express notable surprise when they were shown that there was no treat inside of the mitten.

Cutts and Ceci (1988) interpreted this as evidence that early reproductive memory can be unconscious; preverbal children may have encoded the event in a manner that did not allow the trace to survive the 4-month retention interval at a conscious level. However, unconscious reproductive memory was clearly evidenced by the observed touching behavior. In a departure from the typical finding of enhanced remembering when a reactivation of cues is employed, Cutts and Ceci had a subgroup of children for whom a natural reactivation existed: These children's families went on vacation before the entire 15-day acquisition period was complete, and even though they had already fully acquired the behavior in question, they were given additional trials when they returned from vacation several weeks later. Although these children had already achieved the training level needed to establish criterion before they went on vacation, as a group their recall was no better than that of the children who were given all of their experiences with the cues during the original training period, that is, they did not receive cue reactivation. If reactivation, is a concept of any breadth and generality, it may need to be elaborated to account for such failures of spaced training facilitation.

Surprising results that are relevant to discussion of both day-to-day reproductive memory and to the issue of infantile amnesia have been produced in a study by Perris et al. (1990). In their work, 1.5- and 2.5-year-olds who had participated in a study of auditory localization at the age of 6.5 months returned to the laboratory. In the original experiment, which each infant had experienced on only one occasion, the infants reached in the light or dark for an object emitting a sound. In the follow-up at the later ages, the infants were reintroduced to the procedure in the dark, and were given uninstructed and then instructed trials in which they were to find the sounding object. The results showed that among 2.5-year-olds, those who had experienced the task in infancy reached out and grasped the object significantly more often than a control group without infancy experience. Among the 1.5-year-olds, however, this was not the case. A possible reason for this age-related difference is that subjects in the younger control group showed significantly higher reaching and grasping behaviors than their older counterparts, making the effects of infancy experience more difficult to assess in the 1.5 year olds.

An additional result of interest is the fact that experienced 2.5-year-olds appeared to be less startled during the task, and were more likely to remain in the testing situation than children in the control group. An important point about this methodology is that reinstatement of the entire context of the early experience, as opposed to simply selective cues within it, is presumed to be required in order for the early memory to be brought forth.

In line with the reactivation effects achieved by Rovee-Collier and her colleagues, the findings of Perris et al. indicate the efficacy of reminders in infancy. In this case, the shaking of the same rattle used in the original experiment just prior to the beginning of the test trial acted as a reminder of the infancy experience. The essential conclusion drawn by the researchers from their data is that 2.5-year-olds remembered aspects of an experience that they had encountered only once at the age of 6 months. The successful retention of a single experience at such a young age is unprecedented in the literature, and provides a promising glimpse of what new methodologies may uncover about unconscious recollections from early infancy.

CONCLUSION

In summary, both human and animal research with infants has converged on the finding that the young information-processing system is quite efficient and capable of prolonged retention almost from the very beginning, provided that care is taken to supply contextual reminders. Although the study of contextual factors among human infants has been a recent phenomenon, it is one that has been prevalent among animal researchers for some time. Thus, there has been a shift in the way psychologists have approached children's memory. While many continue to depict memory as a singular system that operates with identical efficiency across all contexts (you either have a good memory or a bad one), others have started to view the memory system as highly sensitive to context. These latter researchers are searching for boundary conditions that define the contextual limits on processing efficiency. The assumption in this work is that memory operates with varying degrees of efficiency, depending on both intrinsic factors (e.g., the structure of the knowledge base involved; the affective reactions during exposure) and extrinsic factors (e.g., the physical attributes of the training context). Elsewhere, we have concluded from an analysis of a mnemonist named "Bubbles P." and other mnemonists that,

In contrast to the presumed singularity of memory, our case study, along with more systematic research we have been carrying out with representative samples of children and adults, suggest that memory is extremely context-sensitive. Being good at remembering one kind of material has little implication for being good at remembering another kind: Recalling dance steps, for example, has no predictive implication for recalling digits. In a sense, the case of memory is but a special instance of the more general case of context-specificity in all of cognition. All forms of cognitive performance—from the most basic processes involved in encoding and scanning to the more molar forms of problem solving—are responsive to contextual alterations. (Ceci, DeSimone, & Johnson, in press)

An important commonality in several of the studies mentioned in this chapter that were able to document memory among very young children is that it may find expression at different levels of consciousness. Often, memory is evidenced in these studies by so-called *implicit* measures, such as willingness to remain in the test situation, or a higher incidence of playing with some object that was used during the familiarization phase, or increased touching of some target feature (e..g, Cutts & Ceci, 1988). These are taken as indications of memory even when some explicit target response is not made. A similar dissociation is seen with adult amnesiacs. For example, they may not remember having been told a joke just minutes before, yet they will nevertheless laugh less at it the second time it is told to them (Jacoby, 1982), and may reflect differences in activities occurring during encoding and retrieval. Roediger and Blaxton (1987) suggested that implicit memory reflects data-driven processes that are triggered by the immediate perceptual information available during encoding and retrieval. Explicit memory, in contrast, entails conceptually driven processing that includes strategic activities such as elaboration and organization. Infant memory may be more readily expressed implicitly because of limitations in their conceptually driven processes, coupled with their greater dependence on activities which rely on contextual aspects of the situation, or data-driven processes.

Taking a somewhat different interpretive tack, Howe and Courage (in press) argued that infantile amnesia occurs during the first 2 years because of the emerging self-representational system. For example, a 1-year-old girl who stares at herself in the mirror after having had lipstick smeared on her nose will not try to remove it by wiping her nose, even though she is able to identify other object locations in a mirror. This is taken by some to indicate that the child of this age has not developed a full self-representation yet. Therefore, it is pointless to expect the child of this age to recall autobiographical details, that is, personally experienced events that are tied to time and place. Very young children may exhibit amnesia for these early experiences because they do not possess a self-concept that can serve to organize the cues and facilitate recall. In contrast, the implicit events that have been reported in the literature may be less autobiographical. For example, it does not require a self-representation to touch the mitten that was associated with a treat. One possible problem with this interpretation is that there is some evidence that even 1-year-olds do have developed self-concepts, and their mirror difficulties may reflect their undeveloped sense of optical flow.

Some of the research reviewed in this chapter may seem at odds with the phenomenon of infantile amnesia discussed earlier. For example, it was argued that the expression of long-term explicit memory would seem to require reinstatement of the context and/or the use of reactivation treatments. And yet, children do not forget everything they ever experienced or learned before the age of five, especially those things that are continually reexperienced (e.g., brushing teeth, tying shoes), but even things that are not (e.g., upon return to a vacation cottage they will correctly seek out a favorite hiding place they had not seen since their trip the prior year). More subtle indicators of memory may be demonstrated without contextual aids, but the durability and/or specificity of these memories may be compromised as a result. It may be that the dysfunctional behaviors that might manifest themselves later are the remnants of those more implicit or subtle early memories.

Finally, we return to the everyday implications of the infantile amnesia research. Was the 42-month-old girl in Moran v. Commonwealth of Massachusetts correct in her claim that she had been abused around the age of 20 months by her maternal grandmother and grandfather? The sad fact is that the research is still a very long way from being useful in such forensic contexts because at present there is no agreement on the boundary conditions for various contextual effects, nor on the efficacy of concepts like reactivation and self-representational systems. For instance, since the little girl was attending therapy with her mother—an alleged survivor of childhood sexual abuse herself—and had repeatedly been exposed to her mother's claims that the grandparents had abused her when she was young, and had been asked if her grandparents had ever done such "things" to her, one could claim that this repeated questioning served to reinstate the context of the abuse, and hence to aid in its retrieval. On the other hand, there are obvious dangers inherent in situations in which adults persistently suggest scenarios to young children (for review, see Ceci & Bruck, in press). From the present literature, we know that early experiences can get into the memory system and are capable of being later retrieved, at least implicitly. We also know that the phenomenon is too restless to be nailed down with any precision. For instance, the former view that maintained that experiences prior to the age of three or four could not later be consciously retrieved is now known to be false. Usher and Neisser (in press) have shown that adults are capable of retrieving some experiences that occurred to them as early as 24 months (birth of a sibling), while other experiences may not get into permanent memory until the age of 3 (moving house) and still others may not get in before the age of 4 (death of a relative). These researchers ruled out alternative explanations for their subjects' recall, such as rehearsal or the subsequent provision of descriptions by family members. If they are correct, then this shows that the offset for infantile amnesia is sliding, depending on the salience of the event, the degree of its rehearsal, and probably a host of additional factors.

ACKNOWLEDGMENTS

Portions of this research were supported by a grant from the National Institute of Child Health and Human Development, RO1 HD24775-01A2 and were written when the first author was in Brazil on a Senior Fullbright-Hayes Award. Address correspondence to Steve Ceci at HDFS, Cornell University, Ithaca, NY 14853.

REFERENCES

Abrams, R. (1984). *Attorney General's Report to the New York State Legislature.* May 13. Albany: State Documents Office.

Accredolo, L. P. (1979). Laboratory versus home: The effect of environment on the 9-month infant's choice of spatial reference system. *Developmental Psychology, 15,* 666–667.

Bartlett, F. C. (1932). *Remembering: A study in experimental and social psychology.* London, UK: Cambridge University Press.

Bauer, P. J., & Mandler, J. (1990). Remembering what happened next: Very young children's recall of event sequences. In R. Fivush & J. Hundson (Eds.), *Knowing and remembering in young children* (pp. 9–29). New York: Cambridge University Press.

Campbell, B. A., & Jaynes, J. (1966). Reinstatement. *Psychological Review, 73,* 478–480.

Canfield, R., & Haith, M. M. (1991). Young infants' visual expectations. *Developmental Psychology, 27,* 198–298.

Ceci, S. J. (1990). *On Intelligence . . . more or less: A bioecological treatise on intellectual development.* Englewood Cliffs, NJ: Prentice Hall.

Ceci, S. J., & Bronfenbrenner, U. (1985). Don't forget to take the cupcakes out of the oven: Strategic time-monitoring, prospective memory, and context. *Child Development, 56,* 175–190.

Ceci, S. J., & Bruck, M. (in press). The suggestibility of the child witness: A historical review and synthesis. *Psychological Bulletin.*

Ceci, S. J., & Leichtman, M. (1992). Memory, cognition, and learning: Developmental and ecological considerations. In S. Sigalowitz & I. Rapin (Eds.), *Handbook of Neuropsychology* (pp. 223–239). Holland: Elsevier.

Ceci, S. J., DeSimone, M., & Johnson, S. (in press). Memory in context: A case study of "Bubbles P.", a gifted but uneven memorizer. In D. J. Herrmann, C. McEvoy, & H. Weingarten (Eds.), *Memory improvement and memory theory.* Hillsdale, NJ: Lawrence Erlbaum Associates.

Ceci, S. J., Ross, D. F., & Toglia, M. P. (1987). Age differences in suggestibility: Psycholegal implications. *Journal of Experimental Psychology: General, 116,* 38–49.

Cutts, K., & Ceci, S. J. (1988, August). *Long-term recall of experiences in the first two years of life: Memory for cheerios of cheerio memory?.* Paper presented at the meeting of the American Psychological Association. Atlanta, GA.

Ebbinghaus, H. (1885). *Über das Gedächtnis* (Translated by H. Ruger & C. Bussenius Memory, 1913). New York: Teacher's College, Columbia University.

Fagan, J. F. (1979). The origins of facial pattern recognition. In M. Bornstein & W. Kessen (Eds.), *Psychological development from infancy: Image to intention* (pp. 1–27). Hillsdale, NJ: Lawrence Erlbaum Associates.

Fagan, J. F. (1971). Infants' recognition memory for a series of visual stimuli. *Journal of Experimental Child Psychology, 14,* 453–476.

Freud, S. (1963). Introductory lectures on psychoanalysis. In J. Strachey (Ed.), *The standard edition of the complete psychological works of Sigmund Freud* (Vol. 15). London: Hogarth Press.

Goodman, G. S. (1980). Picture memory: How the action schema affects retention. *Cognitive Psychology, 12,* 473–495.

Goodman, G. S., & Reed, L. (1986). Age differences in eyewitness testimony. *Law and Human Behavior, 10,* 317–332.

Goodman, G. S., Rudy, R., Bottoms, B., & Aman, C. (1990). Children's concerns and memory: Issues of ecological validity in the study of children's eyewitness testimony. In R. Fivush & J. Hudson (Eds.), *Knowing and remembering in young children* (pp. 249–284). New York: Cambridge University Press.

Haroutunian, V., & Riccio, D. (1979). Drug-induced arousal and the effectiveness of CS exposure in the reinstatement of memory. *Behavioral and Neural Biology, 26,* 115–120.

Hill, W. L., Borovsky, D., & Rovee-Collier, C. (1988). Continuities in infant development. *Developmental Psychobiology, 21,* 43–62.

Howe, M. L., & Courage, M. L. (in press). On resolving the enigma of infantile amnesia. *Psychological Bulletin.*

Hudson, J. (1990). The emergence of autobiographical memory in mother-child conversation. In R. Fivush & J. Hudson (Eds.), *Knowing and remembering in young children* (pp. 166–196). New York: Cambridge University Press.

Istomina, M. (1975). The development of voluntary memory in preschool aged children. *Soviet Psychology, 13,* 5–64.

Jacoby, L. (1982). Knowing and remembering: Some parallels between Korsakoff patients and normals. In L. Cermak (Ed.), *Human memory and amnesia* (pp. 97–122). Hillsdale, NJ: Lawrence Erlbaum Associates.

Kessen, W., & Nelson, K. (1978). What the child brings to language. In B. Z. Presseisen, D. Goldstein, & M. H. Apperl (Eds.), *Topics in cognitive development* (Vol. 2). New York: Plenum.

King, M., & Yuille, J. (1987). Suggestibility of the child witness. In S. J. Ceci, D. Ross, & M. Toglia (Eds.), *Children's eyewitness memory* (pp. 24–35) New York: Springer-Verlag.

LeCompte, G. K., & Gratch, G. (1972). Violation of a rule as a method of diagnosing infants' levels of object concept. *Child Development, 43,* 385–396.

Lewkowicz, D. J., & Turkowitz, G. (1980). Cross-modal equivalency in early infancy. *Developmental Psychology, 16,* 597–607.

Lickliter, R., & Hellewell, T. B. (1992). Contextual determinants of auditory learning in Bobwhite quail embryos and hatchlings. *Developmental Psychobiology, 17,* 17–31.

Mandler, J. (1984). Representation and recall in infancy. In M. Moscovitch (Ed.), *Infant memory: Its relation to normal and pathological memory in humans and other animals* (pp. 75–110). New York: Plenum.

McDonough, L., & Mandler, J. (1990, April). *Very long-term recall in two-year-olds.* Poster presented at the International Conference on Infant Studies, Montreal.

Meltzoff, A. (1988a). Infant imitation after a 1-week delay: Long-term memory for novel acts and multiple stimuli. *Developmental Psychology, 24,* 470–476.

Meltzoff, A. (1988b). Infant imitation and memory: Nine-month-olds in immediate and deferred tests. *Child Development, 59,* 217–225.

Moran v. Commonwealth of Massachusetts. (1988).

Myers, N. A., Clifton, R., & Clarkson, M. G. (1987). When they were young: Almost-threes remember two years ago. *Infant Behavior and Development, 10,* 123–132.

Nelson, K., & Ross, G. (1980). The generalities and specifics of long-term memory in infants and young children. In M. Perlmutter (Ed.), *Children's memory: New directions for child development* (pp. 87–202). San Francisco: Freeman.

Ornstein, P. A., Gordon, B., & Larus, D. M. (1992). Children's memory for a personally experienced event: Implications for testimony. *Applied Cognitive Psychology, 6,* 49–60.

Ornstein, P. A., Larus, D. M., & Clubb, P. (1992). Understanding children's testimony: Implications of research on the development of memory. In R. Vasta (Ed.), *Annals of child development Vol. 8.* London: Jessica Kingsley Publishers.

Perris, E., Myers, N. A., & Clifton, R. K. (1990). Long term memory for a single infancy experience. *Child Development, 61,* 1796–1807.

Roediger, H. L. III, & Blaxton, T. A. (1987). Retrieval modes produce dissociations in memory for surface information. In D. S. Gorfein & R. R. Hoffman (Eds.), *Memory and cognitive processes: The Ebbinghaus centennial conference* (pp. 349–379). Hillsdale, NJ: Lawrence Erlbaum Associates.

Rovee-Collier, C. (1990). The "memory system" of prelinguistic infants. In A. Diamond (Ed.), *The development and neural bases of higher cognitive functions.* New York: Oxford University Press.

Rovee-Collier, C. (1984). Ontogeny of learning and memory in infancy. In R. Kail & N. E. Spear (Eds.), *Comparative perspectives on the development of memory* (pp. 103–134). Hillsdale, NJ: Lawrence Erlbaum Associates.

Rovee-Collier, & Hayne, (1987). Reactivation of infant memory: Implications for cognitive development. In H. W. Reese (Ed.), *Advances in child development and behavior* (Vol. 20, pp. 185–238). New York: Academic Press.

Rudy, J. W., & Cheatle, M. D. (1979). Ontogeny of associative learning: Acquisition of odor aversions by neonatal rats. In N. E. Spear & B. A. Campbell (Eds.), *Ontogeny of learning and memory* (pp. 157–188). Hillsdale, NJ: Lawrence Erlbaum Associates.

Smith, G. J., & Spear, N. E. (1978). Effects of home environment on withholding behaviors and conditioning in infant and neonatal rats. *Science, 202,* 327–329.

Smith, G. J., & Spear, N. E. (1980). Facilitation of odor aversion conditioning by the presence of conspecifics. *Behavioral Neurobehavioral Biology, 28,* 491–495.

Smith, P., & Hull, G. (1984). Five-month-old infant recall and utilization of temporal organization. *Journal of Experimental Child Psychology, 38,* 400–414.

Spear, N. E. (1979a). Experimental analysis of infantile amnesia. In J. Kihlstrom & F. Evans (Eds.), *Functional disorders of memory.* Hillsdale, NJ: Lawrence Erlbaum Associates.

Spear, N. E. (1979b). Memory storage factors leading to infantile amnesia. In G. Bower (Ed.), *The psychology of learning and memory* (Vol. 13). New York: Academic Press.

Spear, N. E., Kraemer, P. J., Molina, J. C., & Smoller, D. E. (1988). Developmental changes in learning and memory: Infantile disposition for unitization. In J. Delacour & J. C. S. Levy (Eds.), *Systems with learning and memory abilities* (pp. 124–148). North Holland: Elsevier.

Spear, N. E., & Kucharski, D. (1984). Ontogenetic differences in stimulus selection during conditioning. In R. V. Kail & N. E. Spear (Eds.), *Comparative perspectives on the development of memory* (pp. 227–252). Hillsdale, NJ: Lawrence Erlbaum Associates.

Spear, N. E., Kucharski, D., & Miller, R. R. (1989). The CS- effect in simple conditioning and stimulus selection during development. *Animal Learning and Behavior, 17,* 70–82.

Spear, N. E., & Molina, J. C. (1987). The role of sensory modality in the ontogeny of stimulus selection. In N. A. Krasnegor, E. M. Blass, M. A. Hofer, & W. P. Smotherman (Eds.), *Perinatal development: A psychobiological perspective* (pp. 85–108). New York: Harcourt Brace Jovanovich.

Spear, N. E., & Parsons, P. J. (1976). Analysis of reactivation treatments: Ontogenetic determinants of alleviated forgetting. In D. L. Medin, W. A. Roberts, & R. T. Davis (Eds.), *Processes of animal memory* (pp. 135–166). Hillsdale, NJ: Lawrence Erlbaum Associates.

Squire, L., & Cohen, N. (1984). Human memory and amnesia. In G. Lynch, J. McGaugh, & N. Weinberger (Eds.), *Neurobiology of learning and memory* (pp. 3–64). New York: Guilford Press.

Strauss, M. S., & Cohen, L. B. (1980, April). *Infant immediate and delayed memory for perceptual dimensions.* Paper presented at the International conference on Infant Studies, New Haven, CT.

Topinka, C. V., & Steinberg, B. (1978, March). *Visual recognition memory in 3½ and 7½ month-old infants.* Paper presented at the International Conference on Infant Studies, Providence, RI.

Usher, J., & Neisser, U. (in press). Childhood amnesia in the recall of four target events. *Journal of Experimental Psychology: General.*

Werner, J. S., & Perlmutter, M. (1979). Development of visual memory in infants. In H. W. Reese & L. Lipsitt (Eds.), *Advances in child development and behavior* (Vol. 14). New York: Academic Press.

8 PIFS: Everyday Cognition Goes to School

Joseph Walters
Tina Blythe
Noel White
Harvard Graduate School of Education

Consider the variety of talents that school children have. Imagine, for example, three very different, but typical, 7th graders: Ariel, Beth, and Chris. Ariel is a straight-A student who loves school despite her discomfort when other children call her the "teacher's pet." Teachers expect something less from Beth, an average student and an excellent gymnast, whose friends consider her the leader of their group. Chris has a passion for playing the violin, and, according to her music teacher, has enormous potential. Because she is self-conscious about her talent, however, she never plays the violin in school.

Are the middle-school experiences of typical students such as these a harbinger of adult life? Can we predict that Ariel will continue to be the most successful one from this group? To a large extent, the answer is no: School performance is not a completely reliable predictor of adult experience. To illustrate, imagine the biographies of these children carried forward into adulthood. Beth found a niche in management. Although she struggled through a night-school MBA program, once she landed a job, she was quickly promoted through the ranks, becoming the youngest vice president in her Fortune-500 company. Chris gave up the violin in 10th grade because it took too much time away from her social life. However, her disciplined approach to music served her well in adulthood: She found a challenging job as a software engineer for a company that designs database applications. Finally, Ariel did quite well throughout her entire school career, straight through graduate school and right up to the time when she began working on her thesis in psychology. At that point, she discovered that no one had ever asked her to go beyond analyzing other people's work or to create something unique and interesting of her own. Faced with this new challenge, she lost interest in psychology and turned instead to a vocation that demanded an originality of a different sort—working with children. Ariel became a creative and demanding high school English teacher.

These three imaginary case studies highlight the inherent problem in rating Ariel, and students like her, as *bright* or *talented.* Such a rating misses something very important about the capabilities of the other two children. To capture the differences in talents among children—especially individuals as different as Ariel, Beth, and Chris, we need a theory that treats intelligence in a pluralistic fashion. Today there are a number of such theories (Ceci, 1990; Feldman, 1980; Sternberg, 1988). In this chapter, we examine one developed by Howard Gardner called Multiple Intelligences theory (Gardner, 1983) in an effort to understand how traditional schooling fails to address the various capabilities of students.

We begin with a brief outline of the theory of Multiple Intelligences and use this theory to analyze the unique environment of formal schooling. We then turn to an investigation undertaken at Harvard Project Zero and Yale University called Practical Intelligence for School (PIFS) that draws on this pluralistic theory of intelligence in order to understand more completely the relationships among traditional schooling, individual students, and the society in which both students and schools exist. In the concluding section we describe how our thinking about the theory of Multiple Intelligence has changed given these experiences with the PIFS Project.

THE THEORY OF MULTIPLE INTELLIGENCES

In the theory of Multiple Intelligences (MI), an intelligence is defined as an ability to solve problems or fashion products that are valued in a cultural context. The theory identifies seven distinct competencies, each of which constitutes an intelligence: linguistic; logical-mathematical; spatial; bodily-kinesthetic; musical; interpersonal; and intrapersonal. The data to support such a claim emerged from several fields, including research on human development and neurology, studies of special populations (including autistic children, idiot savants, and child prodigies), findings from cross-cultural research, and evidence offered by the study of evolution (Gardner, 1983).

These various aptitudes or intelligences are universal; that is, they are found in all humans (with only rare exceptions) and in all cultures. The intelligences are also tightly bound to specific cultural contexts. For example, the linguistic skill (or intelligence) is universal and its development in children is surprisingly constant across cultures. At the same time, it can be manifested in many different ways in adults of different cultures: the use of spoken words to tell a story, or the use of signs in American Sign Language to give directions, or the use of letters to produce a secret code. These activities also occur in a cultural context: The story might be a parable told by a religious leader; the directions might be instructions for finding one's way through a new city; the code an initiation to a secret society.

Because the intelligence and its deployment in a social context are interrelated, the attempt to describe a native intelligence apart from the context in which it is used can be misleading. For example, an individual who excels at public speaking must use that linguistic intelligence in a different way to tell stories effectively. Someone who can give directions succinctly may be clumsy at debate. In each case, the native intelligence may be said to remain unaffected in some sense, while the individual's ability to mobilize the intelligence effectively changes.

Complex tasks—such as conducting a debate, giving directions, telling a story—involve more than one intelligence. Giving directions combines linguistic and spatial intelligences. Telling a story involves linguistic and interpersonal intelligences; and debate draws on logical intelligence. Given this need for a mix of abilities, individuals with similar linguistic intelligence have different profiles of success on tasks that demand combinations of abilities. Because the interaction of abilities and contexts is a central part of MI theory, one cannot meaningfully speak of the pure "intelligence," without also specifying a particular context in which the competence is manifested.

The importance of context has implications for the evaluation of an individual's abilities. To evaluate any of these intelligences, one must pose problems in the rich context of a specific domain. For example, the musical intelligence is brought into play when an individual composes a melody. In most cases, music composition presupposes some degree of formal training in music, but it has been shown that novices can compose melodies and solve simple problems of composition when they use a computer (Scripp, Meyaard, & Davidson, 1989). The music composer, either novice or expert, confronts a series of genuine musical problems, and in solving them, deploys the musical intelligence. In contrast, this musical ability is not revealed in any complete sense when one is asked to talk about musical problems or to answer questions about music on a multiple-choice test.

In addition to this focus on the contexts in which individuals work, MI Theory suggests that the interpersonal and intrapersonal intelligences play particularly important roles in deploying the other intelligences effectively. Interpersonal intelligence defines our ability to understand and share the meanings, motives, intentions of one another. This skill is critical in nearly every adult role—in work relationships, in parenting, and in all forms of group interactions.

The intrapersonal intelligence—that sense we have of ourselves, our abilities, liabilities, and aspirations—is the most speculative of the seven intelligences and the most difficult to see in action. The intrapersonal intelligence produces an internal portrait of our strengths and weaknesses from which we make decisions and predictions when we enter new and challenging situations. For example, if one were introduced to several new people at the same time, one's intrapersonal intelligence might produce the judgment, "I would prefer to socialize with these people individually, since that is what I do best, instead of speaking to them as a group." A judgment of this sort is made from an analysis of personal strengths and weaknesses based on previous experience.

As with the other intelligences, each of us has this ability to make personal discriminations to some degree, but the sophistication of this intelligence varies from one individual to another. For instance, imagine a person who has innate strength in the realm of mathematics but who fails to recognize the presence (or the importance) of this capability. Or picture an individual who is highly talented in linguistic and logical arenas and assumes that he or she has an equivalent talent in an unrelated area—say for managing people. Students like Ariel can fall into this category. Based on her success at school tasks, which require logical and linguistic skills, Ariel may assume that other demanding tasks, such as finding a job, will pose no difficulty for her. These judgments are the result of drawing the tempting but fallacious inference that strength in one intelligence implies strength in another.

In summary, the theory of Multiple Intelligences pluralizes the concept intelligence, describing the mind as a combination of several distinct abilities that manifest themselves in specific domains and in specific social settings. These core abilities can be seen at work only as the individual manipulates the materials, solves the problems, and fashions the products of a given domain. Each of us possesses a unique profile of these abilities, a profile that is at once dependent on human endowment and on the entire history of one's social and physical interactions within particular cultural contexts. Personal knowledge of this profile and of its relation to others around us is important for effective use of these abilities. Since the theory of Multiple Intelligences focuses on the deployment of these intelligences in a variety of cultural settings (in contrast to studying them exclusively through psychological tests), the theory has a strong "everyday" flavor to it. This tie between the intelligences and the environment in which they are manifested proves to be essential in understanding the everyday environment of school.

USING THE MI FRAMEWORK TO EXAMINE THE CONTEXT OF SCHOOL

Analyzing the Academic Curriculum

Our analysis of school from the perspective of Multiple Intelligences begins by examining the academic curriculum. The kinds of skills and knowledge required by the school curriculum can differ in striking ways from those used in the world outside of school. Resnick (1987) depicts the school context as one of generalized learning, symbol manipulation, and pure mentation. For example, arithmetic is presented as a collection of algorithms that are performed on numbers without units of measure or context. Reading is often reduced to decoding printed symbols into vocalized utterances. Writing might focus on grammar and spelling; and history seldom emphasizes the important relationships that obtain among the names, places, and dates that the students memorize. This milieu contrasts sharply with the outside world of situation-specific competencies, contextualized reasoning, and tool manipulation.

The theory of Multiple Intelligences asserts that the context of a problem can have a marked impact on how one finds a solution (see also Ceci, 1990; Cole & Scribner, 1974; Wason & Johnson-Laird, 1976). In the terms of MI, the school curriculum—with its textbooks, practice problems, chapter tests, and so on—sets the linguistic and the logical-mathematical intelligences in a highly symbolic and abstract context. Those who work best with these intelligences and in this context—students like Ariel—are most likely to succeed in school. Others whose strengths are not well tapped in this context (like Beth and Chris) are more likely to have difficulty in school. Our analysis of school will focus mainly on this mismatch between abilities and context as a source of students' difficulties in school, although many other factors, including environmental factors—a destructive home life, limited resources available at school, racial discrimination—also may feed into a particular child's difficulties.

Providing us with an example of the disparity between the school context and an out-of-school type of problem, Resnick describes children's responses to a problem that occurs most typically outside of school: If an ice cream cone costs 60 cents and you have in hand a quarter, a dime and 2 pennies, how much more do you need to buy the cone? In a real-world setting, this question would probably lead to searching through the pocket for "round change"—a quarter if necessary, or dimes and nickels. In a school context, however, the problem might be interpreted as an invitation to calculate. Resnick describes children working in small groups on this problem who unwittingly gave evidence to the lack of fit between the real-world solution involving pockets of coins and the school solution involving numerical calculation. One girl, who resembles Ariel, quickly did the subtraction and announced, "The answer is twenty-three." This went unheeded by other children who interpreted the problem in a more real-world way. The group eventually figured out that, ignoring the two pennies in hand, an additional quarter was needed to buy the cone. But the girl who had originally worked out the subtraction did not want to give up her solution. In an attempt to convince the group, she announced, "We could get the change from out of the quarter," implying that they would then use just twenty-three cents of it (Resnick, 1987)!

A second illustration is taken from the investigation of children's understanding of temperature. In an experiment conducted by Strauss, Stavy, and Orpaz (1977), children were asked a number of questions about what happens when one mixes hot water with cold water. They answered several of the questions correctly: hot and hot makes hot; a lot of hot and a little cold makes warmer, and so on. They could accurately fill in worksheets that presented these various conditions with beakers and thermometers. All descriptions were qualitative, *hot* or *warm* or *cold* water.

A second group was given virtually the same problems, except that numerals were included on the worksheets that the students filled out. *Cold* water was water at 10° (Celsius) and *hot* water was 60°. This group was also given equipment—beakers, thermometers, and a water supply. This second group answered the questions incorrectly. Ten-degree water and 10° water when mixed produce 20° water. One student, after predicting 20° water in the mixture, measured the result at 10° and replaced the thermometer with a second one. After getting the same measurement with the second thermometer, he exchanged it for a third one. Upon making a third reading of 10°, he turned to the observer and said, "Some days nothing works right!" (Strauss et al., 1977)

Here the role of context is made clear through a clever experimental design. Again, the impulse to add—presumably learned in the abstract world of school—is so strong that it overrides all other considerations and distorts all other knowledge of the problem space, including immediate tactile experience and any semblance of common sense. How might Multiple Intelligences characterize this anecdote? Including the numbers in the second situation effectively changed the nature of the problem: Children were no longer working in a tactile, common-sense context; instead, they were working in a school context. MI theory argues that the problem in this different context taps different abilities, thereby changing students' success in solving the problem. Too often, school fails to make this connection. Students are asked to struggle with problems that might be solved more easily and accurately if approached in a different context.

Analyzing the Social Context of School

In addition to considering the academic setting, the theory of Multiple Intelligences, with its definition of the personal intelligences, provides a framework for analyzing the social context of school. From the point of view of students, the social issues, challenges, and problems of school are equally or more salient than the academic ones. Ask typical 6th graders why they like going to school (and many of them do!), and they often will say, "Because my friends are there" (Goldman, Krechevksy, Meyaard, & Gardner, 1988). Unfortunately, this social context is often ignored by the curriculum. Social skills are rarely taught explicitly. Skill in this area—aside from proper behavior—often goes unrewarded. Generally, socialization among students is carefully controlled, and adults often ignore social difficulties, except in the most serious cases.

From the MI point of view, again, there is a mismatch between school and other social settings. Children like Beth, whose interpersonal skills serve them well in other settings such as clubs, scouts, or informal peer groups, will find that those skills may not serve them as handily in school. Although an understanding of teachers, other students, administrators, and such is important for success in school, these social skills do not constitute the overt curriculum. The context of solving problems outside of school, which often requires that one work well in a group, does not necessarily play a role in school success.

School From the Intrapersonal Perspective

What individuals understand or fail to understand about themselves—their intrapersonal skill—can have a marked impact on their willingness to take on and solve unfamiliar problems successfully. Not only does school do little to develop this skill, it complicates the task by offering students a view of themselves that is often quite different from the one they generate in their experiences outside of school. In the school setting symbols are paramount and individualized performance is the operating standard. But outside of school, individuals manipulate tools not symbols and group work is often the operating standard. How does one create an accurate and complete picture of self, relying on such information?

A sophisticated intrapersonal intelligence will yield a differentiated internal picture that takes into account varying situations and conflicting data. Indeed, some children seem to possess this rich sense of self almost innately: They understand the mismatch between their abilities in school and out of school and how to deploy those abilities differently in each situation. We suspect that these children will be, generally speaking, more successful in school. Other children, especially those with a less sophisticated intrapersonal intelligence, will have a harder time. They will misjudge their unique blend of strengths and weaknesses. And they will fail to draw on their strengths effectively, especially those skills that function quite well outside of school, as they confront school's more challenging problems. It is this second group—children who are "at risk" because they fail to match their personal profile of skills to the demands of school—that is the focus of the PIFS Project.

THE "PRACTICAL INTELLIGENCE FOR SCHOOL" PROJECT (PIFS)

There are two possible approaches to helping these at-risk students. The first is to change the structure of school itself—to broaden the goals and methods of public education in such a way as to nurture the intelligences of all students. This Olympian task would require a dramatic change in fundamental assumptions about education; and it would rest on agreement among educators on precisely what those assumptions are. This kind of change is well beyond the scope of a small research project.

The second alternative is to change the students by preparing them to deal with school as it exists. The Practical Intelligence for School (PIFS) project adopts this second approach. Sponsored by the McDonnell Foundation, PIFS is a 6-year research collaboration between Robert J. Sternberg at Yale University and Howard Gardner at Harvard Project Zero at the Harvard Graduate School of Education. The project has devised and tested methods and materials that middle school teachers can use to help students develop *practical intelligence*—those special skills that are important in school but are often ignored by the formal curriculum.

The PIFS Interviews

In an effort to determine more precisely which particular skills—or practical intelligences—students deploy as they negotiate their way through school, we conducted a series of in-depth interviews with 50 5th- and 6th-grade students. We asked these students about their study habits, their grades, the roles teachers play in their education, how they interact with peers, and more generally how they understand school itself. The interviews reveal that children differ markedly in their understanding of school and of their place in it. Some students display an awareness of themselves as learners who make active use of a variety of strategies and resources for learning. In contrast, other students lack these strategies and respond passively to their school experiences. We describe the first group as students with "high PIFS" profiles and the second as students with "low PIFS" profiles (Goldman et al., 1988).

To illustrate, students with the high PIFS profile gave more detailed answers to our questions. They offered multiple solutions to problems, identified several educational goals for themselves, and listed different ways of achieving those goals. They were more precise in their determination of personal strengths and weaknesses. Moreover, they could articulate a number of strategies for overcoming weaknesses. Children with the low PIFS profile, on the other hand, gave simple, unelaborated responses. When asked how they handled their weaknesses, they often answered in terms of a single, global strategy: "Try harder, study more." One student explained, "Everything helps a little, but not that much."

With respect to the intrapersonal intelligence, the high PIFS children were more likely to connect the purpose of their activities in school to their long-term personal goals. Low PIFS children resorted to punitive explanations for school. "You go to school because you have to" or "If you don't do the homework, you get in trouble."

When the interview shifted to a discussion of activities outside of school, however, the students with low PIFS profiles responded as competently as the high PIFS students. Almost all low PIFS students had at least one area in which they seemed to feel knowledgeable and capable. For example, one student provided few details when asked why school was difficult for him; but he became animated and articulate when talking about music or architecture, two of his favorite topics. Other low PIFS children talked enthusiastically about their ambitions to be attorneys, actors, dancers, singers, or athletes, although they rarely saw any connections between these long-term goals and their work in school.

The high PIFS students, then, understand what is expected of them in school. They know how to satisfy these expectations while at the same time fulfilling their own needs. The students with the low PIFS profiles lack both a feel for the expectations of school and the sense of self-efficacy that would enable them to satisfy those expectations appropriately. They see school as a place where one responds passively because active work toward solutions to problems is pointless, where knowledgeable adults are not resources for overcoming obstacles, where the purpose of their activity is mysterious and unknowable. If these students are to succeed in school, then they must be equipped with the tools necessary for doing so: a sense of self-efficacy, an explicit understanding of the relevance of school to life, and the ability to marshal a variety of resources, both personal and external, to overcome obstacles.

Description of the PIFS-Infused Units

The Yale research team developed a self-contained curriculum that addresses these issues. The Project Zero team produced units which can be integrated with curricula already used in classrooms. In this section we focus on these *infused* units.

Our analysis of the school setting based on the theory of MI together with the results of our PIFS interviews gave rise to the main foci at the core of each unit: (a) context, and (b) the personal intelligences. First, the modules were designed to help students understand the context surrounding specific school problems through an emphasis on the purposes of various school tasks. Students are asked to think explicitly about the relationship between school and their extracurricular lives—how school tasks are similar to and different from what students do in the real world. In this way the PIFS modules expand the narrow context of school problems.

Second, PIFS modules help students to develop intrapersonal intelligence and to draw on that faculty as they work on school problems. Each module stresses the importance of reflection and provides opportunities for students to reflect on their own strengths, weaknesses, and experiences. The modules also formalize the role of interpersonal intelligence in school and strive to develop interpersonal skills by presenting activities to which cooperative learning is integral.

Each of the 10 infused units addresses a particular issue or topic which both teachers and students identified as problematic. (Thirteen units were developed originally. For a variety of reasons, three of them were not tested adequately in the classroom. This discussion reports the data obtained from the other 10 units.)

TABLE 8.1
PIFS Infused Units

Reading and Writing Units
Understanding Fiction: Reading Between the Lines--focuses on strategies for discerning the deeper, symbolic meaning in stories. **Choosing a Project**--helps students to choose and plan school projects that are well-suited to both class requirements and their personal interests.
Mathematics Units
Word Problems--helps students to focus on the process of solving math problems, the variety of potential problem-solving strategies, and their own most effective problem-solving techniques. **Finding the Right Math Tools**--leads students to become familiar with a range of math resources and to develop strategies for using them efficiently.
Social Studies Units
Using a Variety of Resources--fosters students' awareness of the diversity of available resources, as well as of their own personal resources. **Making Research Reports Your Own**--helps students to consider alternative report formats which both discourage copying and promote a deeper understanding of new information.
General Units
Organizing and Presenting Your Work--helps students to develop personally appropriate way of organizing and presenting their work. **Taking Notes**--fosters the development of more efficient note-taking skills. **Why Go to School**--leads students to consider the broader connections between school and their personal lives, both in the short and long runs. **Discovering Your Learning Profile**--encourages students to identify their own special array of abilities, particularly nonacademic abilities, and to draw on them when confronting difficult situations.

Six of these focus on problems tied to specific content areas: two in language arts, two in social studies, and two in mathematics. The remaining 4 address issues and skills that obtain across domains (see Table 8.1).

To illustrate how these units function in classrooms, we describe in some detail one content-specific unit, "Understanding Fiction: Reading Between the Lines," and one general unit, "Why Go to School." Like the other modules in the 3 content areas, "Understanding Fiction" was developed to service a specific problem. Children in middle school have difficulty understanding fiction. They may understand the words and sentences, but they often miss the deeper, symbolic meanings in fiction. They don't read between the lines.

To help them with this task, the module begins by placing the activity of reading fiction within the larger context of typical school reading. The students examine textbooks—the reading matter that they associate most closely with school—and compare them to novels. (Their initial characterizations are fairly simple: A textbook is "fact" whereas fiction is "fake.") Through further discussion, the class moves toward a more complex understanding of fiction—to see that it, too, contains information of a sort. However, this information is presented in a markedly different way from the information in a textbook. This understanding leads to the conclusion that the familiar way of reading textbooks is inappropriate for novels. A different kind of reading is needed for fiction.

The unit then presents lessons that help students to see the similarities between everyday activities with which they are familiar, and the still-foreign process of reading for nonliteral meaning. Through discussions of commonplace symbols (such as flags and advertising logos), students realize that in fact they are experienced decoders. In brainstorm and reflection activities, students learn that the context in which a symbol appears creates the web of associations which give the symbol meaning. Students also come to appreciate individual differences: Because everyone has different associations with particular symbols, the meanings each person attaches to them will be different. Ultimately, students try their hand at creating symbols within original commercials and short stories. With classmates, they practice interpreting, critiquing, and defending interpretations.

This module is typical of many in the set. It is grounded in concrete examples and connected to experiences students have had outside of school. It relies on group discussions to elicit concepts. And eventually it offers students opportunities to assume responsibility for applying the knowledge in new situations to solve specific problems.

The "Why Go To School" unit helps students to situate school-as-a-whole in the broader context of their present lives and future goals (both short and long term). A series of activities and discussions leads students to a clearer understanding both of the school as a social system and of themselves as learners.

Students reflect on the place of school in the context of larger society by examining graphs that compare the job prospects of high school graduates and nongraduates. They interview adult professionals to discover how school affected their choices and opportunities. To situate their current experiences in a more immediate context, students analyze the school curriculum, comparing what they are studying currently with what they studied in the earlier grades. In discussing the differences between the two, the students come to a clearer understanding of the purposes of school and how they change over time.

Other activities encourage intrapersonal intelligence. Students talk about those aspects of school that they enjoy the most or like the least. They discuss why it might be important to "put up with" some problems, and develop solutions or compromises to ameliorate others. They give careful thought to their own goals and interests, and then discuss how school supports or neglects those personal projects. Each student has the opportunity to design a curriculum that she feels would help her the most.

The intensive small group work that accompanies this unit fosters the interpersonal skills necessary to working cooperatively with others. Both the process and the content of the activities encourage students to recognize points of view other than their own. And understanding the diverse needs and goals of the many students who attend one school is a crucial first step in appreciating school itself as a complicated social institution.

The Study

All PIFS units were developed in collaboration with several experienced teachers who served as consultants to the project. As the units were developed, they were tested in three middle schools and then revised based on comments from teachers and observations of researchers.

We also developed evaluation measures for each unit. Each evaluation includes items of three different types. *Definitional* questions gauge the students' understanding of the issues addressed in the PIFS unit—can the student describe why the issue is important, or relate it to their goals? *Task* items sample the actual skills targeted in the unit. Finally, a *meta-task* component asks students to step back and reflect on the nature of the process or skills involved in a particular task.

After the units and the evaluation measures were developed and tested, we conducted a study to evaluate their effectiveness in changing students' practical intelligences. This study was implemented over a 2-month period during the spring semester of the 1989–90 school year in 6 Boston-area public elementary and middle schools. Nine teachers chose one or two of the PIFS units that they felt would be most relevant to their classes. All students in these experimental classes and their control counterparts (matched for grade and socioeconomic background) were given a pretest.

The experimental teachers then met with researchers to discuss the units and to tailor the core lessons of each unit to their current curriculum. During the first and second weeks of implementation, these teachers presented the core lessons from chosen units. Over each of the next 4 to 6 weeks, they conducted one follow-up lesson that reinforced the key concepts of the unit. All students in the experimental and control classes were given a posttest approximately 2 weeks after the final follow-up lesson in the experimental class.

Results

In addition to the quantitative information obtained through the pre- and posttests, we documented the effectiveness of the units using several qualitative sources—classroom observations, teacher observations, interviews with teachers and students. By all accounts, the PIFS curriculum units proved successful in the classroom. Particularly encouraging was the overwhelmingly positive response of teachers who found that the PIFS units provided them with a systematic way to address the needs of students who are not skillful learners. These teachers also reported that working through the PIFS lessons gave them an opportunity to sharpen their focus on and philosophy about the teaching of particular issues.

Drawing on observations and interviews, we also determined that students benefited from the PIFS materials. They developed techniques for handling important school tasks that had been difficult for them in the past. They began to recognize the complexities both in the topics they were studying and in their own learning styles. Finally, virtually all students—particularly those who otherwise were not usually engaged constructively in class—enjoyed the PIFS activities.

The quantitative data from pre- and posttests support these results. On all units, the experimental group's scores improved from pre- to posttest. To determine the significance of these improvements, the Mann-Whitney U Test was applied to compare the difference scores of the experimental group (posttest minus pretest) for each of the 10 units with the difference scores of the control group (see Table 8.2). The experimental group's scores showed significant gains on 5 units, and insignificant gains relative to the control group's scores on 5 units. Suspecting that the small sample size might make it difficult to show significance, the results of these last five units were also analyzed using the Fisher Exact Probability Test. By this additional measure, the relative improvement of the experimental scores for 2 units, "Discovering Your Learning Profile" and "Why Go To School," approached significance.

Discussion

The PIFS skills that we identified as important can be taught effectively. An examination of the characteristics of the most successful units gives some indications about the circumstances under which students best learn these skills. All units that yielded significant gains for the experimental group were domain-

TABLE 8.2
Results for the Experimental Groups from the Mann-Whitney U Test for Statistical Significance

Units Fully Implemented	*z-Scores*	*Significance*
Understanding fiction: Reading between the lines	3.04/2.58	yes/yes*
Choosing a project	3.06	yes
Math tools	1.17	no
Word problems	3.44	yes
Using a variety of resources	3.17	yes
Making research reports your own	3.60	yes
Organizing and presenting your work	.29	no
Taking notes	.35/1.15	no/no*
Why go to school	1.10	no
Discovering your learning profile	1.45	no

*The two scores are for the two different classes in which the unit was implemented.

specific. Of the five units that did not produce significant gains, all but one (an interesting exception that is discussed later) were general, or not tied to a specific domain. This underscores the appropriateness of an infusion approach to teaching practical skills. The domain-specific approach ensures that PIFS lessons have obvious relevance to students' school work. For example, students learn about PIFS skills appropriate to project work as they do actual school projects. They examine abilities involved in solving word problems as they work on daily math assignments. They use stories assigned for English reading to learn about symbolism in fiction.

In addition, these units encouraged students to make use of extracurricular interests and experiences in carrying out specific school tasks. By incorporating extracurricular interests and experiences, the PIFS lessons provide motivation and help students to recognize already existing skills that can be important to school success.

Another important factor in the success of these units may have been that each of them was implemented in conjunction with another unit. The teachers who taught combinations of units adopted PIFS strategies more fully than other teachers. These units were given more time, and became more thoroughly infused into classroom routine, allowing for activities from the different units to reinforce each other.

Two of the general units, "Why Go To School" and "Discovering Your Learning Profile," included many features of the domain-specific units, with an important exception: They did not focus on specific tasks, but rather on students' attitudes toward and feelings about school. Both only approached significance on the Fisher Test. This is not to say, however, that these units were without effect; we simply were not able to demonstrate that effect statistically. We know, for example, that following the "Why Go To School" unit, one class created a special once-a-week activities period in which students could pursue independent projects (with the teacher's blessing). Another class designed a special "conversation class," which would feature the development of social skills. An 8th-grade class used the module to help them with the challenging task of deciding what courses they would take at the high school the following year. While qualitative indicators such as these suggest that these topics can be taught, growth in these areas in a short period of time is difficult to document quantitatively.

The less successful test results for the other two general units, "Taking Notes" and "Organizing and Presenting Your Work," may be related to the fact that these units closely resemble traditional study skills curricula. Again, comments from teachers and students indicated that these units were well-received in the classrooms. However, the teachers had taught the topics before, so the experimental groups' responses on the pretests were already fairly high. Their scores did improve on the posttests, but the improvement was slight. Furthermore, most non-PIFS classes also teach students directly about note-taking and organization.

As might be expected, then, the control groups' scores for both of these units also went up slightly from pre- to posttests, making the effects of the PIFS instruction in the experimental classes difficult to interpret.

"Math Tools," the only content-area unit that failed to produce significant gains in the experimental group, reveals another trait of the PIFS materials—their highly compelling nature. We noticed that 28 out of 30 students in the *control* group for this unit improved on the posttest. Curious about this effect, we contacted the control group teacher, who explained that she had shared the pretest with the math teacher on her teaching team. He had been fascinated by the questions, and began to incorporate some of the ideas and issues into his class discussions. Although we are delighted that the materials have this kind of resonance with teachers, and we are happy with the progress of the students in the control group, there is the distinct possibility that this unexpected coaching from the math teacher spoiled the experiment with this unit.

CONCLUDING REMARKS

The PIFS project has a number of messages. For the students, the message is that, with the help of resources within themselves and others, they *can* figure out how to handle school successfully—how it relates to the outside world and how they as students differ from their classmates.

Also, there is a message here for the teachers. The PIFS units model the connections that teachers must make between school and the outside world. Students need direct instruction in matters that require interpersonal and intrapersonal skill. The PIFS modules encourage teachers to do what many of them already do naturally—make a connection between the abstractions of school and everyday cognition, especially those skills of interpersonal and intrapersonal intelligence that children bring to school.

Our experimentation with the PIFS units suggests that carefully designed curriculum materials can help in both regards. These materials can help students prepare for the challenges of school by placing school in the broader context of the practical world. They can help teachers by making the connection more explicit and by structuring the discussions and considerations that students will find helpful.

Finally, our work with PIFS has a message for psychologists. It indicates that an analysis of learning and cognition that focuses on domains of human endeavor can have important payoffs when applied to schooling. This perspective in turn reaffirms the view taken by MI, that thinking can best be understood in the context of daily events.

What we have learned here can be applied in other contexts as well. Indeed, in working with the PIFS project we have come to think of the outside world as actually a number of different spheres, each requiring its own combination of practical intelligences. These might include such things as work, family, intimate relationships, and leisure.

Each of these domains conceivably could have its own practical intelligence units—practical intelligences for families, for managers, for skiing, and so on. Following the model of our Practical Intelligence For School, these units would help individuals determine their unique blend of intelligences and to compare that blend with the demands of a particular domain in a particular cultural context. In a sense, the many self-help books that are so popular approach precisely these issues for the lay audience. But just as PIFS has tried to systematize the insights into the domain of school, these new PIFS units would take a systematic approach to their respective domains.

These new Practical Intelligence units would focus attention on individual differences and the application of personal strengths to identified problems. And just as our PIFS units try to expand beyond a reliance on the verbal realm, the new units would encourage development of the nonverbal intelligences as well. Beyond these design similarities, however, the new practical intelligence units would be as different from PIFS in content as school is from work.

Therefore, whatever the domain, we must be wary of the Ariel phenomenon—identifying performers on the basis of a single, high-profile ability while ignoring the multiplicity of skills and talents which are needed to complete any task. Instead, we must remember to look broadly, to identify individuals with unique and unexpected sets of skills. We hope that curricula such as PIFS may be able to help both the students and the teachers to understand themselves and one another more fully and, from this, to develop a deeper grasp of the very complex relationship among individual talents, school performance, and everyday life.

ACKNOWLEDGMENTS

The work described in this paper was supported by a grant from the James S. McDonnell Foundation. We would like to thank the students, teachers, and administrators of the PIFS Project. We are also grateful to Howard Gardner and Mara Krechevsky for many helpful comments on earlier drafts.

REFERENCES

Ceci, S. (1990). *Intelligence—more or less: A bio-ecological treatise on intellectual development.* Englewood Cliffs, NJ: Prentice Hall.

Cole, M., & Scribner, S. (1974). *Culture and thought: A psychological introduction.* New York: Wiley.

Feldman, D. (1980). *Beyond universals in cognitive development.* Norwood, NJ: Ablex.

Gardner, H. (1983). *Frames of mind: The theory of multiple intelligences*. New York: Basic.

Goldman, J., Krechevsky, M., Meyaard, J., & Gardner, H. (1988). *A developmental study of children's practical intelligence for school* (Harvard Project Zero Technical Report).

Resnick, L. (1987). Learning in school and out. *Educational Researcher, 16*(9), 13–20.

Scripp, L., Meyaard, J., & Davidson, L. (1989). Discerning musical development. In H. Gardner & D. Perkins (Eds.), *Art, mind, and education* (pp. 75–88). Urbana, IL: University of Illinois.

Sternberg, R. (1988). *The triarchic mind: A new theory of human intelligence*. New York: Viking.

Strauss, S., Stavy, R., & Orpaz, N. (1977). *The child's development of the concept of temperature*. Unpublished manuscript. Tel-Aviv University.

Wason, P., & Johnson-Laird, P. (1976). *Psychology of reasoning: Structure and content*. London: Batsford.

9 Adolescents' Thinking About Emotions and Risk-Taking

Alexander W. Siegel
Paula Cuccaro
Jeffrey T. Parsons
Julie Wall
University of Houston

Armin D. Weinberg
Baylor College of Medicine

> *The "teens" are emotionally unstable and pathic. It is the age of natural inebriation without the need of intoxicants, which made Plato define youth as spiritual drunkenness. It is a natural impulse to experience hot and perfervid psychic states, and is characterized by emotionalism.*
>
> —(G. Stanley Hall, 1904)

> *Of all stages of life, adolescence is the most difficult to describe. Any generalization about teenagers immediately calls forth an opposite one. Teenagers are maddeningly self-centered, yet capable of impressive feats of altruism. Their attention wanders like a butterfly, yet they can spend hours concentrating on seemingly pointless involvements. They are often lazy and rude, yet when you least expect it, they can be loving and helpful.*
>
> *This unpredictability, this shifting from black to white and from hot to cold is what adolescence is all about.*
>
> —(Csikszentmihalyi & Larson, 1984, p. xiii)

We live in a period of heightened *adolescent-consciousness*. Not only are there a lot of adolescents around, they seem to have both more disposable income and more problems than ever before, and we still don't understand them. They drive us nuts and they still remain (as they have since Plato's time) enigmatic.

The puzzling nature of adolescence is nearly *magnetic* in its attraction for researchers and social commentators alike. For the last few years, we too have been drawn to the magnet of adolescence. In what follows, we (a) provide a brief glimpse of the everyday subjective world of adolescents (i.e., their *umwelt*), (b) describe some of our research on adolescents' thinking about emotions and risk taking, and (c) discuss aspects of both adolescent and adult egocentrism as heuristics for describing and explaining how adolescents think about their physical and social worlds.

ADOLESCENT MYTHS AND REALITIES

Just as the moods of teenagers oscillate, so research on adolescents has peaked at various times in this century from G. Stanley Hall's romantic and recapitulationistic *Adolescence* (1904) to Hartshorne, May, and Maller's work on "character" (1929), to Coleman's (1961) work on delinquency and the youth culture, and to Piaget's descriptions of formal operations (e.g., Flavell, 1963). In the 1980s, once again, increased public attention and scientific research was directed toward adolescence. This focus was visible in the number of new journals dedicated to adolescence, a new society (the Society for Research in Adolescence), the great variety of books written on how to *parent* adolescents, the national legislation passed, the Presidential Task Forces commissioned, and heightened awareness via the media of a variety of teenage problems including gangs, pregnancy, drugs, and suicide.

As in much of psychology, G. Stanley Hall got there first (see Siegel & White, 1982). In his massive two-volume *Adolescence: Its psychology and its relations to anthropology, sociology, sex, crime, religion and education,* Hall (1904) argued that adolescence is a "second birth" and as such entails a universal and inevitable period of *sturm und drang* (storm and stress), during which individuals are driven by biology and characterized by extreme and unpredictable mood swings.

The popular myth that adolescence is universally and inevitably a time of emotional storm and stress and social turmoil is, in large measure, attributable to Hall's *Adolescence*. Although some prefer to hold on to the belief that all adolescents are in continual emotional turmoil and at constant war with their parents (e.g., Blos, 1941, 1979), many others regard adolescence as a period of relatively smooth transition for most individuals (e.g., Offer, Ostrov, & Howard, 1981; Savin-Williams & Demo, 1984). Coleman (1978) has analyzed these two apparently contradictory points of view in some detail. Opposing the classical view of *sturm und drang* is the empirical perspective, which asserts that the teenage years are far more stable and peaceful than the storm and stress theorists would have us believe. In fact, Siddique and D'Arcy (1984) have suggested that the classical and empirical points of view may not be contradictory at all; it depends on your focus. For some teenagers adolescence is, indeed, characterized by storm and stress; for others, the majority, it is relatively peaceful. While the debate isn't over (this one cannot be decided completely on the basis of *data*), at the very least, the myth of universal storm and stress is itself undergoing serious and intense storm and stress.

Most theorists would agree, however, that adolescents do face a number of situations that must be negotiated and which involve a considerable amount of emotional energy. While adolescence involves new and difficult social, psychological, and biological negotiations, from the empirical perspective (e.g., Coleman, 1978) teenagers are a lot more together than we give them credit for. Indeed, it can be argued that adolescence is more traumatic for parents than it is for their teenagers (Davis, 1940; Steinberg, 1989).

The qualitative changes that occur in many adolescents' thinking have been a topic of research since Piaget became popular. We have impressive documentation of the differences between adolescent thought and that of younger children, and we know that teenagers develop the capacities to engage in abstract, "rational," "scientific," "hypothetico-deductive," and "allocentric" thinking (Keating, 1980). Early work focused on adolescent thinking about nonsocial objects and relationships—propositional and symbolic logic, how they performed on analogies and syllogisms, how they solved geometry problems, and the like. More recently, as the phrase *social cognition* became more popular, the nature of the research shifted towards how adolescents viewed social objects and relationships (Selman, 1980), and how their knowledge of and judgments about themselves and others became more complex, differentiated, and integrated (Hill & Palmquist, 1978).

For example, some researchers (e.g., Davis, 1940; Grotevant & Cooper, 1985; Hill & Holmbeck, 1987) have noted that children initiate more arguments in families as they enter adolescence. Skeptics and parents have typically seen this as a trait expression of rebelliousness, that is, they impute increased argumentativeness to adolescents (rather than seeing it as an index of logical abilities and rhetorical skill). The *rebellious* interpretation stems from—and perpetuates—the myth of adolescence as a period of inherent crisis and turmoil (Hall, 1904). Later, research on (and public fascination with) delinquency and the psychopathic personality (e.g., McCord & McCord, 1956) led to the portrayals of adolescent rebellion by James Dean and Marlon Brando in the late 1950s. By the 1960s, rebellion was no longer viewed as a behavior limited to individuals, but was assumed to have spawned an entire deviant youth culture (Coleman, 1961, 1978). Part of this readiness to assume the worst about young people may have something to do with the short (and inaccurate) memories of adults, ("When I was a boy I walked barefoot in the snow to school, eight miles uphill, both ways"). In matters of style, taste, and etiquette, young people of every generation have seemed radically different from adults in how they look and act, and in their music, hairstyles, and clothing.

But careful research (e.g., Youniss & Smollar, 1985) indicates that young people's basic values differ very little from those of their parents. It is a mistake to confuse most adolescents' enthusiasm for trying on new identities and enjoying moderate amounts of outrageous behavior with genuine hostility to parental and societal standards. Being boorish and testing the limits are time-honored ways in which young people, beginning in infancy, move toward accepting, rather than rejecting, parental values. As Napier and Whitaker (1985) pointed out, "The sword is forged in the family and tempered in the peer group" (p. 38). It does little good and much harm to think of adolescence as a time of rebellion, crisis, pathology, deviation, or social epidemic (Csikszentmihalyi & Larson, 1984). Twenty-five years ago, Douvan and Adelson (1966) pointed out that while interaction between parents and teenagers is far from smooth, psychologists and writers may have overestimated the amount of stress actually occurring between parents and teenagers. As Offer (1969) argued (see also, Offer & Offer, 1975; Offer, Ostrov, & Howard, 1981) from his extensive research on normal and disturbed adolescents:

> Normal adolescents are not in the throes of turmoil. The vast majority function well, enjoy good relationships with their families and friends, and accept the values of the larger society. In addition, most report having adapted without undue conflict to the bodily changes and emerging sexuality brought on by puberty. (pp. 116–117)

In line with Offer's perspective, it can readily be posited that teenagers initiate and engage in more arguments because they become better arguers.

In effect, adolescence is defined in terms of relatively profound reorganizations in children's cognitive, social, and physiological worlds. Puberty involves changes in physical size, strength and proportion, hormonal and CNS-regulated functions, secondary sex characteristics (such as body hair, acne, voice register changes), and changes in coordination. These physical changes are apparent to the adolescents themselves, their friends, and their parents. Social changes include—among other things—individuation, the increasing importance of peers, and status-ambiguity. While a lot of changes in a fairly short period of time need not be debilitating or throw the young person into crisis, it can be confusing (leading to what Erikson [1951] described as "identity diffusion").

We have argued elsewhere (see Siegel, Bisanz, & Bisanz, 1983; White & Siegel, 1976, 1984) that learning—and by extension, socialization, and cognitive development—is largely a process of "becoming unconfused." To the extent that this metaphor has heuristic utility, perhaps the cartoon (see Fig. 9.1) provides a glimpse of what the *umwelt* of a teenager is like. Situational uncertainties, expectational ambiguities, and schooling-career-money contingencies engender strong emotions and, not infrequently, provoke conflicts. So do concern about peers, homework, AIDS, and diet (adolescents and adults alike get emotional about the dangers of cholesterol and the benefits of broccoli).

FIG. 9.1. *Umwelt* of an adolescent. Copyright © (1989) by David Sipress, New American Library. Reprinted by permission.

ADOLESCENT THINKING ABOUT EMOTIONS

Most of our descriptions of adolescents center on the intensity, extremeness, and lability of their emotional expression, but only recently have psychologists begun to study directly these emotions. Csikszentmihalyi and Larson (1984; Larson & Richards, 1989) have explored the phenomenological (i.e., emotional and experiential) and social-ecological aspects of being adolescent, using the Experience Sampling Method (ESM). In their research, adolescents carry pagers, are "beeped" randomly during every waking 2-hour block and asked to report (by filling out a brief form on-the-spot) their situation—where they are, whom they are with, what they are doing, and how they feel. From a large number of these subjective (but quite reliable) self-reports, the authors found that the emotions experienced by adolescents were similar to those reported by adults, but that teenagers experienced more extreme and intense mood states and more rapid fluctuations between states.

These findings are consonant with parents' and teachers' everyday experience. The attention of teenagers wanders—from being incredibly on-task in something in which they are interested to being in another reality (Earth calling teenager).

As Bill Cosby has deftly described, lapses of memory and sanity occur—"I don't know" ("why I shaved off half my hair and colored the rest blue"). Agony and ecstasy are but moments apart. Extremes of experience are *desiderata* and are reflected in the hyperbole of adolescents' descriptions of events: "Awesome" turns out to be one JND above good; "the pits" is one JND below minimally adequate; "excellent" apparently means anything better than acceptable; nearly everything is "boring." Teenagers apparently want and need to do sensation seeking (within imposed limits); they need to test their own limits (hopefully, only infrequently to the breaking point)—and those of others. (As some of our research is beginning to indicate, alcohol and drug use is more frequently seen as a means to these ends than as conscious risk-taking).

Curiously, little or no empirical research has looked directly at these phenomenologically valid, but "soft," intuitions and converted them into valid consensual knowledge. In order to do so, some relatively unorthodox and different methodologies will have to be used. American developmental psychology "discovered" Piaget in the 1960s in part because many felt that we had pretty much milked dry the old paradigms, and in part because we became less fussy about where and how we got interesting data. Apparently we have reached the same point in adolescent thinking, as a number of researchers are becoming increasingly frustrated by paradigmatic and procedural limitations of standardized tests and $2 \times 2 \times 2$ experimental designs.

Some healthy, speculative rethinking and reconceptualizing of both the ecology and phenomenology of adolescents is in the making. As Barker and his colleagues (e.g., Barker, 1968; Barker & Wright, 1955; Schoggen, 1991) have elegantly shown, a full-blooded understanding of adolescent thought and behavior must rest on a detailed analysis of the adolescent's ecology. Such objective accounts that tabulate the number of same- and different-age and gender friends contacted over the course of a week, or list the number of emotional outbursts on a weekday vs. a weekend, are true to the outside reality of adolescence, that is, true to the changes that an observer can detect in their behavior. What these accounts leave out is the inside or subjective reality—what it is like to be a teenager, what teenagers do and think in their everyday lives, how they feel about themselves and their situations.

TOWARD A STRUCTURE OF ADOLESCENT EMOTIONS

As a modest first step in this enterprise, we began by examining the content and structure of adolescent emotions, as these are expressed by teenagers in their everyday language. Words were used to index (or reflect) emotions, since words (speech) frequently mediate between phenomenological experience and behavioral expression (Wertsch, 1985). Further, there is some evidence indicating long-term stability of emotional expression in adolescents. For example, Savin-Williams and Demo (1984) examined self-feelings by giving adolescents beepers and randomly paging them over a period of 1 week per year between the 7th and 10th grades. Individuals were given 20 positive and 20 negative words and indicated those that represented their feelings at that particular place and time. Results indicated that self-feelings remain relatively stable from year to year. Thus, although adolescents appeared to be unstable, volatile persons over the short run (moment to moment or day to day), over a more extended period of time their feelings remained relatively stable.

Our own first study was exploratory. We wanted to develop an initial model of the structure of adolescent emotions, and assess the extent to which emotion words clustered together. (*Model* is used in its informal sense, as a rough mapping to adolescents' internal reality, *not* as a formal device for making predictions, with the obligatory arrows and boxes of contemporary cognitive science). Prior to the formal study, three exploratory pilot studies were conducted. The first was an interview conducted with one 16-year-old female. A structured interview and an open-ended discussion involved the individual's phenomenological experience—time spent in different states, labels used for those states, feelings experienced in these different states, moving between emotional states, etc. This initial model was explored further in a second pilot study involving three other 16-year old females, who were given in a group a structured interview that followed up more precisely on hunches we had developed during the first pilot study. The three girls discussed their own emotional experiences, first in an open group discussion and subsequently using a picture of the model itself.

The level of sophistication in theorizing about relationships *among* emotions is fairly low-level. Emotions have been categorized using two different theoretical approaches, either in terms of basic universal emotions or in terms of dimensions across which the emotions are related (Bretherton, Fritz, Zahn-Waxler, & Ridgeway, 1986). The first approach (e.g., Ekman, 1973; Izard, 1971) posits a limited number of basic emotions, such as anger, happiness, sadness, and fear, each of which has separate processing systems that deal with qualitatively different information and motivate qualitatively different behaviors. In the dimensional approach—the one adopted in this study—emotion is defined in terms of at least two orthogonal dimensions: hedonic tone and activation.

Before the turn of the century, Wilhelm Wundt (Blumenthal, 1975) proposed a theory of affect which included three dimensions: pleasant vs. unpleasant, high arousal vs. low arousal, and concentrated attention vs. relaxed attention. Schlosberg (1954) proposed rather similar dimensions in his theory of affective behavior: pleasantness–unpleasantness, high activation–low activation, and attention–rejection. Osgood, Suci, and Tannenbaum (1957) proposed one of the most widely known dimensional theories of affect, based on their research with the semantic differential. The dimensions that characterized emotions were: good–

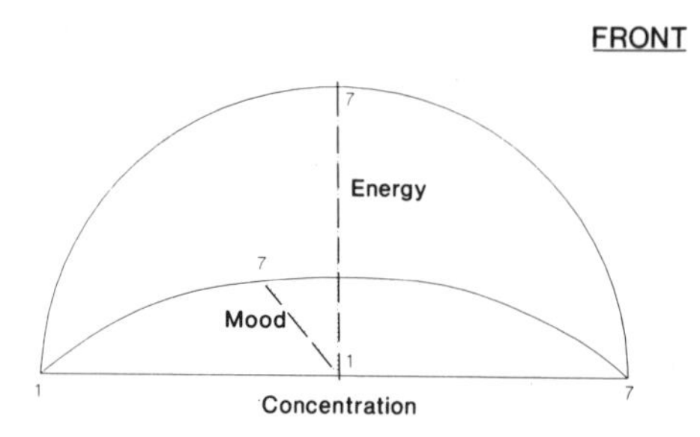

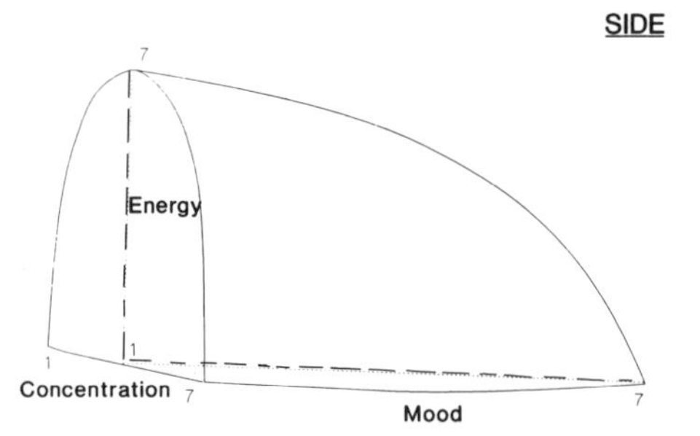

FIG. 9.2. Model of the structure of emotions in adolescent thinking.

bad, active–passive, and strong–weak. Curiously enough, our initial model—based on the pilot data—included dimensions most similar to Wundt's.

The initial model in Fig. 9.2 can be described as a bandshell having three dimensions: mood (very negative to very positive), energy (very low to very high), and concentration (very low to very high). The points on the model are based on an ordinal scale (1–7) only. As one moves to the edges, the intensity of the state becomes more extreme. Each set of three coordinates (energy, concentration, mood) characterizes a state involving the three dimensions. This model clearly made sense to the girls and expressed the way they reported experiencing the states and moving in and out of emotional states. Please note that we are not claiming (nor do we mean to imply) that adolescents (or anyone else) naturally think about their emotional states using this metric—if you ask me how I am feeling this afternoon I will tell you that "I'm exhausted," not "1, 2, 3"! For instance, [6, 2, 1] constitutes an emotional state with a high level of energy, a low level of concentration, and very negative mood—possibly described as fear, panic, or hysteria. In contrast, at [6, 2, 6] there is also a high level of energy, a low level of concentration, but a very positive mood—possibly described as euphoric. The point of total concentration [7], focused completely on the self, has mood and energy scale coordinates of [4,4]. The individual in this state appears "out of it" to others, although she feels totally aware and in control. The girls could place emotion words onto the initial model with little difficulty, once they became accustomed to considering the three dimensions simultaneously.

They could also determine which place they would be most likely to move to next within the initial model's space from any other place in the model's space. For example, on the mood dimension, there appeared to be a bridge between 3 and 5 which could be moved across readily. But when the individual had reached a more extreme state (1 or 7) the bridge no longer existed.

The questionnaire used for the formal study was based on the pilot data and on this preliminary model. The questionnaire incorporated the three dimensions of the model as well as three other constructs that appeared to be promising theoretically or methodologically: frequency, intensity, and self-control. Seventy-eight 11th and 12th grade high school students and 189 undergraduates (mainly freshman and sophomores, but excluding anyone over 22) participated in the formal study. The high school sample was quite homogeneous—caucasian, high GPA, aged 16 or 17; the college sample was more heterogeneous—mixed ethnicity, GPA spread, aged 18–22. Thirty-two words were selected from a larger set of 80 words used in a third pilot study; subjects were asked to rate the frequency with which they used the words in their everyday discourse. Each student rated the 32 selected emotion words on 7-point Likert-type scales. Six different scales were included, one for each dimension studied (mood, concentration, energy) and one for each of the three additional constructs of interest (intensity, frequency and self-control). Subjects were asked to rate each word (on all six scales). On all scales except frequency, subjects rated how they would feel (e.g., how positive or negative, how energetic, etc.) if they were experiencing that emotion; on the frequency scale they rated how often they experienced that emotion. Of the 32 words, nine were dropped because more than 60% of the respondents rated the words as being used relatively infrequently.

The rating data on the remaining 21 words were subjected to two kinds of analyses. First, factor analyses were conducted to explore the underlying structure of the data. Second, mean values for each word for the mood, energy, and concentration dimensions were calculated (across subjects) and plotted to give a more readily interpretable visual picture of the extent to which the data fit the proposed model.

For both samples, although males and females differed, the factor structures before and after controlling for gender displayed only slight differences. Exploratory alpha factor analysis with varimax rotation was used, with separate analyses conducted for each sample for each dimension. In both samples, the underlying factors included a combination of the mood and energy dimensions in our model, and the concentration dimension was not revealed as a factor either alone or in combination with other dimensions.

While none of the dimensions had *exactly* the same factor structure in both the high school and college samples, the similarities were striking (as shown in Table 9.1). The factor analyses of the frequency, energy, mood, and self-control constructs yielded similar results in terms of number and order (i.e., strength) of factors for both samples. The concentration construct yielded similar results in

TABLE 9.1
Factor Structures on Each Construct for High School and College Samples

	Sample	
Construct	*High School*	*College*
Frequency	passive positive active negative	passive positive active negative
Energy	passive positive active negative active positive passive negative	passive positive active negative active positive depression
Mood	passive positive active negative turmoil active positive	passive positive turmoil active negative arousal
Concentration	passive positive active negative arousal passive negative	passive positive active negative negative
Intensity	arousal passive positive depression	passive positive passive negative active negative arousal
Self-control	passive positive active negative depression	passive positive active negative arousal passive negative

terms of order but not in terms of number for the two samples. Only the intensity construct was different in both number and order across samples. In sum, high school students and college students appeared to have rather similar emotional structures.

We thought initially that exploratory factor analysis would yield three main factors resembling "mood," "concentration," and "energy." The resulting factor solutions for all constructs, however, were primarily a combination of the mood and energy dimensions of the initial model. The *passive positive* factor appeared on all the constructs for both samples and consisted of such words as "chilled out," "comfortable," "content," "peaceful," "relieved," and "together"; the *active negative* factor showed high loadings for words like "angry," "confused," "crazy," "hyper," "mad," and "mixed up."

These results indicate that the constructs of mood and energy underlie the structure of adolescent emotions. We were not able to find our third factor (and Wundt's)—concentration. Why not? There are lots of possibilities—small *N*s, heterogeneous samples—but most likely the reason rests in the methodology. We thought that adolescents would be better able to concentrate when experiencing certain emotional states, described as "calm," "peaceful," "happy." This is probably the case, but concentration does not constitute one of the aspects of the particular emotion words used. Second, we used factor analysis to see which words hung together. We got factors—not dimensions. A multidimensional scaling task might well result in a concentration dimension, but the factor analytic techniques we used did not. In fact, MDS is a next step, but the factor analysis was needed first.

Given that the factor structures of the high school and college samples were roughly comparable, we wanted to assess how well the 3-dimensional structure of our hypothetical model fit how subjects experienced the stimulus words. To do so, we computed the mean values on each of the model's three dimensions (i.e., mood, energy, concentration) for each of the 21 words (that had been used in the factor analysis) for subjects in both the high school and college samples combined. Using the STATA statistical analysis package, these mean values (presented in Table 9.2) were then portrayed in a 3-dimensional scatterplot (see Fig. 9.3). Because there is no veridical standard against which to compare these mean values (nor against which to compare even individual *maps*), no statistical tests for goodness of fit seem appropriate. An examination of the scatterplot, however, indicates that: (a) the previously identified factors of mood and energy were reproduced; (b) while a concentration "factor" did not fall out of the factor analyses, two rather discrete clusters of words appeared on the graph that reflect low vs. high concentration; (c) visually, the "clusters" of words on the concentration axis were as distinct (if not moreso) as those on the mood and energy axes.

In summary, the factor analyses and scatterplot yielded some interesting, albeit somewhat contradictory, findings. While some may accuse us of going on a fishing expedition in the data, it needs to be remembered that this research was avowedly propadeutic, and not a test of a formal model or set of hypotheses.

> When using FA [factor analysis] the researcher should hold in abeyance well-learned proscriptions against data snooping. It is quite common to use PCA [Principal Components Analysis] as a preliminary extraction technique, followed by one or more of the other procedures, perhaps varying number of factors, communality estimates, and rotational methods with each run. Analysis terminates when the researcher decides on the preferred solution. (Tabachnick & Fidell, 1989, p. 623)

Perhaps most importantly, our pilot studies yielded results at least as interesting and informative as did the formal study. In fact, subsequent to data analyses and informal interviews with other adolescents, it seems more appropriate to conceptualize all three dimensions as bipolar. In this reconceptualization the graphic portrayal of the model looks quite similar to the spherical model initially proposed by Wundt, subsequently rediscovered by Schlosberg and Osgood (Blumenthal, 1970)—and re-rediscovered in our research. We believe that the gains in insight produced through the use of informal and pilot studies are telling us

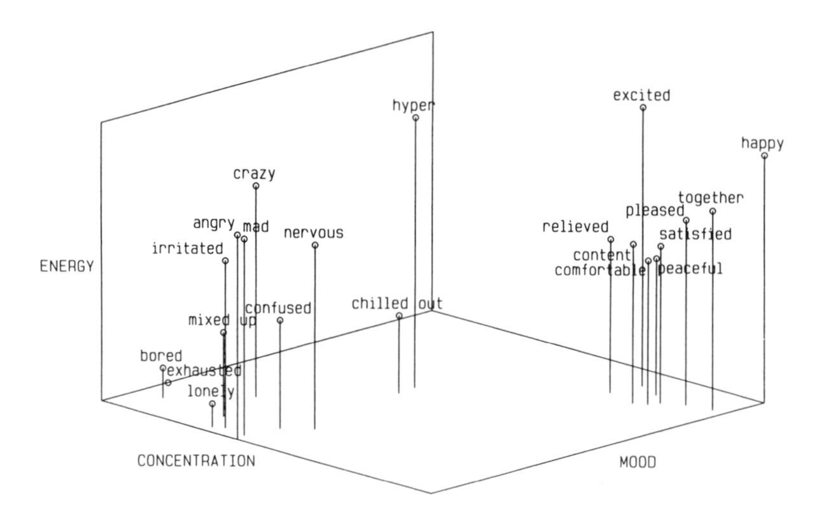

FIG. 9.3. Scatterplot of mean values for the 21 words for all subjects.

something about directions to take in future research, that is, start with the phenomenon, develop a coarse working model, assess its heuristic utility, and move towards more fine-grained analytic procedures to test the validity and generality of a more formal model. We tried to capitalize on this general strategy in the research on risk-taking that we discuss next.

ADOLESCENT THINKING ABOUT RISK-TAKING

The term *risk-taking* frequently refers to behaviors that lead to a high probability of personal harm or injury such as cigarette smoking and alcohol use, as well as other health-related behaviors considered to pose a significant danger to individuals, for example, drug use, unprotected sexual intercourse, driving while drunk, etc. (Furby & Beyth-Marom, 1992). The definition of risk-taking varies depending on who you question—the adolescents themselves, their parents, or research epidemiologists. Adult concepts of adolescent risk-taking behaviors have changed over the past 20 years. The 1970s view considered social forces beyond the adolescents' control to be the main danger to the physical and mental health and welfare of the growing teenager (e.g., Heyneman, 1976; Timpane, Abramowitz, Bobrow, & Pascal, 1976). During the last decade, however, the term *risk-taking* has come to imply that adolescents themselves are largely responsible for their own problematic behaviors (Baumrind, 1987).

Adolescents, however, may well have a different perspective from adults on what specifically does and does not constitute a physical or mental health, or even a social, risk. For example, many AIDS researchers have found that although adolescents *know* the dangers of unprotected sexual intercourse, they fail to adopt preventive behaviors; they fail to apply their knowledge to themselves (e.g., DiClemente, Zorn, & Temoshok, 1986). Adolescents are inundated by the media's antismoking and antidrinking campaigns; they are aware of the negative, long-term health risks, but still continue to engage in these harmful activities.

Cigarette smoking—the largest preventable cause of death in the United States—has been shown to be largely established during adolescence (Best et al., 1984). Drinking alcohol abusively, a widespread health risk in the larger society,

TABLE 9.2
Mean Values for the 21 Words Across all Subjects (N = 219)

	Mood		*Energy*		*Concentration*	
Word	*Mean*	*SD*	*Mean*	*SD*	*Mean*	*SD*
Angry	2.17	1.28	5.21	1.57	3.52	2.04
Bored	2.64	1.29	2.15	1.43	2.43	1.60
Chilled out	4.31	1.39	2.98	1.37	3.48	1.69
Confused	2.71	1.24	3.52	1.48	3.53	2.09
Content	5.59	1.29	4.41	1.32	4.78	1.48
Comfortable	5.65	1.19	4.15	1.36	4.88	1.58
Crazy	3.29	1.73	5.32	1.75	2.86	1.67
Excited	6.06	1.04	6.49	0.81	4.52	1.69
Exhausted	3.04	1.40	1.62	1.16	2.18	1.35
Happy	6.46	0.88	5.96	1.00	5.40	4.06
Hyper	4.54	1.61	6.34	1.15	3.46	1.90
Irritated	2.36	1.16	4.56	1.53	3.26	1.66
Lonely	2.30	1.33	2.03	1.14	3.18	1.83
Mad	2.31	1.14	5.07	1.57	3.48	1.77
Mixed up	2.62	1.24	3.10	1.39	3.05	1.90
Nervous	2.93	1.25	4.85	1.66	3.71	1.92
Peaceful	5.93	1.14	4.02	1.56	4.75	1.66
Pleased	5.89	1.13	4.87	1.28	5.07	1.40
Relieved	5.68	1.14	4.32	1.43	4.49	1.50
Satisfied	5.77	1.18	4.38	1.34	4.91	1.49
Together	5.95	1.27	5.11	1.22	5.28	1.51

also frequently has its roots in adolescence (Rogers, Harris, & Jarmuskewicz, 1987). Further, substance use rates for adolescents increase in early adolescence and consistently rise through late adolescence (Kandel & Logan, 1984). It makes some sense then to understand the development of adolescent perceptions of risk-taking behaviors *prior* to the onset of more-than-occasional use in order to decrease the number of adolescents who move past the experimental phase and eventually develop a smoking habit or a drinking problem.

Parental behaviors *and* attitudes exert a significant influence on the smoking and drinking behaviors of their adolescents (Chassin et al., 1981; McLaughlin, Baer, Burnside, & Polorny, 1985). Parents of adolescents—whether or not they themselves smoke or drink—develop their own perceptions of risk-taking behaviors, which subsequently affect their own adolescents. These parental *perceptions* frequently are more powerful than actual parental *behavior* in shaping adolescent risk-taking behaviors (Newman & Ward, 1989; Nolte, Smith, & O'Rourke, 1983). Consequently, adolescent perceptions of risk-taking should be related to the perceptions held by their parents.

As part of a larger research project designed to examine the role of family variables in young adolescents' attitudes about health-related behaviors, we were able to assess some of these perceptions—the perceived benefits and risks associated with cigarette smoking and drinking in adolescents and their parents. Subsequent to a larger community screening, we were able to recruit and interview 68 families from Victoria, Texas (a southeast Texas city with a population of around 50,000). The target adolescent from each family (23 boys, 45 girls) was either in the 7th (N = 38) or the 9th (N = 30) grade at the time of the interview. Families were selected initially (when the youngsters were in 6th and 8th grade) on the basis of parental smoking status (one or both parents were smokers). The sample was largely White (73.5%), but included families of other ethnicities.

The parents were interviewed together and the adolescent separately. A piece of the interview was constructed to obtain information concerning the parents and the target adolescents' perceptions of the benefits and risks of smoking and drinking. Relevant portions of each interview were coded (with reliabilities in the .85–.95 range) for two variables: (a) frequency—the number of different verbalized risks and benefits; (b) differentiation—the number of different categories into which these responses fell. Examples of benefit categories included: increase good feelings, escape from reality, relaxation, and enhance social situation. Examples of risk categories included: affects others negatively, loss of control, physical damage to self, social damage to self, addiction, and death. The frequency and differentiation scores for smoking and drinking were converted into benefit-to-risk ratios for both parents and adolescents.

More 9th graders reported benefits of smoking (36%) and drinking (40%) than did 7th graders (26%, 29%). Further, more 9th graders than 7th graders gave differentiated benefits for both smoking (37% vs. 26%) and drinking (40% vs.

29%). The frequency ratios and differentiation ratios of 7th graders and their parents were significantly positively correlated for drinking, but not for smoking. In contrast, these scores of 9th graders and their parents were significantly positively correlated for smoking, but not for drinking. Parents who smoked were *less* likely to report risks and *more* likely to report benefits. (Perhaps the parents have learned to justify their habits by attributing to them an inflated number of benefits and denying the consequences). A curious negative modeling effect was found for smoking, but not for drinking—the proportion of adolescents who reported *no* benefits of smoking increased as one or both parents were categorized as smokers; similarly, the proportion of adolescents who reported more risks than benefits of smoking followed the same pattern; *all* of the adolescents who had two parents who smoked reported *no* benefits for smoking *and* a higher number of risks than benefits. In other words, if the parents were conning themselves, the adolescents were not buying it.

The results suggest that adolescents become increasingly aware of the benefits of smoking and drinking. Currently, the antismoking media campaign seems to be working: Among adolescents it is becoming "cool" *not* to smoke. While there has been a media push towards "responsible" drinking, there has been no antidrinking campaign parallel to that for smoking, due in part to the fact that we are a drug-oriented society (Schlaadt & Shannon, 1986) and alcohol is part of the American way of life. Not surprisingly, adolescents think about smoking and drinking differently. Both 7th and 9th graders generated many more benefits for drinking than for smoking. This is consistent with their self-reported *intentions* to engage in risk-taking behavior: 63% of the adolescents (43/68 kids) reported that they intended to drink as adults; only 3% (two kids!), however, reported that they intended to smoke cigarettes as adults. As Wolford and Swisher (1986) have found, *intention* to smoke and drink is the best predictor of engagement in these behaviors 2 years later.

Currently, we are assessing in a large sample of college students (Siegel et al., 1993) and a clinical sample of adolescents (Lavery, Siegel, Cousins, & Rubovits, 1993) their (a) involvement in more than 20 selected "risk" behaviors, (b) perception of the negative consequences of engaging in these behaviors, and (c) perception of the benefits of engaging in these behaviors.

ADOLESCENT AND OTHER KINDS OF EGOCENTRISM

> Youth can not be temperate, in the philosophical sense. Now it is prone to laughter, hearty and perhaps almost convulsive, and is abandoned to pleasure, the field of which ought gradually to widen with perhaps the pain field, although more. There is gaiety, irrepressible levity, an euphoria that overflows in every absurd manifestation of excess of animal spirits, that can not be repressed, that danger and affliction, appeals to responsibility and to the future, can not daunt nor temper. To have a good time is felt to be an inalienable right. (Hall, 1904, p. 77)

Apparently, what is *risk-taking* to adults may not be perceived similarly by adolescents—who may see the same behavior as *experience seeking* (Furby & Beyth-Marom, 1992). The phenomenon seems clear—teenagers engage all too frequently in behavior that has negative consequences—but the semantic descriptor depends on your point of view. By adult standards, it seems reasonable to label a behavior as risk-taking. Even though teenagers know what risky means, and even though they know that a given behavior can have negative consequences, they engage in that behavior anyway. It appears that their decision to take risks is driven by a rationality different from that of the canonical adult. They believe that they are exercising their "inalienable right to have a good time" and that they are bulletproof, i.e., they don't believe anything negative is going to happen to them. This is born out in our preliminary findings that for some 20 behaviors, adolescents' perceived benefits were strongly and positively related to their reported involvements; their perceived risks were negatively and weakly related (Siegel et al., 1993). They do not believe that anything negative is going to happen to *them*.

In 1967, David Elkind published a provocative paper entitled "Egocentrism in Adolescence" (Elkind, 1967), in which he argued that, in effect, too much cognitive development can be bad for you. Elkind argued that with the onset of formal operations, for the first time, the adolescent is able *habitually* to view himself from the perspective of others. This new ability to think about others' thinking, however, is coupled with an inability to distinguish between what is of interest to others and what is of interest to the self. Because the young adolescent is preoccupied with herself—her appearance, her thoughts, her concerns, her fears—she assumes that everyone else is similarly preoccupied with her appearance, thoughts, concerns, and fears. Since adolescence as a distinct social category is a relatively recent cultural invention (Kett, 1977), then, in some larger sense, adolescence can be thought of as a socially constructed thought disorder.

According to Elkind (1967; Elkind & Bowen, 1979), adolescent egocentrism results in two "mini" thought disorders that help to explain (or at least to describe more fully) some of the curious beliefs, contradictions, and behaviors in teenagers. The heuristic utility of the notions of the imaginary audience and the personal fable is well recognized among clinicians, and their reliability and validity have been established by adolescent researchers (Lapsley, 1990; O'Connor & Nikolic, 1990). A major current debate centers on the conceptual linkages between adolescent egocentrism (as indexed by measures of the imaginary audience and the personal fable) and other major adolescent tasks/accomplishments. While Elkind (1967, 1985) argued that the imaginary audience and personal fable (i.e., adolescent egocentrism) arise from the emergence of formal operational thought, Lapsley and his colleagues (Enright, Lapsley & Shukla, 1979; Lapsley & Murphy, 1985) have argued persuasively that the imaginary audience and personal fable both index and elaborate the process of autonomy/individuation. Although the debate about conceptual and causal linkages is unresolved, the existence of the phenomena is not at issue. We believe that these notions have enormous heuristic utility in understanding and studying everyday cognition in adolescents.

The *imaginary audience* is an audience of one's own making. It involves having such a heightened sense of self-consciousness that the teenager imagines that his behavior is the focus of everyone else's attention. For example, when the senior author knocked over his full water glass in a restaurant, he was sure at the time that everyone else heard the sound and was looking at *him* and thinking "What a klutz!" In fact, of course, few people heard it and those few who did, did not care. But at the moment, he was surrounded by an audience *he* had constructed, an *imaginary audience*. We all experience such moments occasionally; but it is pervasive in adolescents. When the senior author's daughter was about 14, she reported that she was worried that *everyone* would notice that she was not wearing designer jeans—at the Rolling Stones' concert in the Astrodome!

This self-consciousness extends into privacy-seeking—for thoughts and language as well as the physical self. Particularly with adolescents, many researchers are frustrated with a frequent parental dilemma—"Where did you go?" "Out." "What did you do?" "Nothing." Although adolescents may be reluctant to talk freely about their own subjective states while alone or with adults—the strange adult with the adolescent in the strange situation—they are much more willing to self-disclose if you talk to them in small groups, a situation much closer to their natural ecology for "sharing."

Closely related, and of more direct interest in our research on risk, is what Elkind calls the *personal fable*. The personal fable centers on the adolescent's egocentric–and erroneous—belief that his or her experiences are *unique*. No one has felt the pain of breaking up—no one could possibly understand what it feels like—as exquisitely as *he* has, he tells his sympathetic mother (even though that is something most people have experienced plenty of times during their adolescent and adult years). "But you don't understand . . ." is yet another way of expressing this "terminal uniqueness."

It is not that great a leap from the feeling of uniqueness to the potentially dangerous belief that "it won't happen to me" or "other people might not be able to drive safely while they're drunk, but I'll be OK." Adolescents believe in their uniqueness and, since their lives are not governed by the principles that govern others, they are immune to the many and varied adverse consequences of risk-taking behaviors. It is only a small inferential step from "I am unique" to "I am exempt." Is it any wonder that many adult-originated intervention strategies aimed at reducing adolescent risk-taking behaviors fail because the target audience does not perceive themselves as vulnerable to any physical or mental danger?

Adults are also, curiously, prone to thinking egocentrically when it comes to risk-taking. Piaget used the term *anthropomorphic* to describe the preoperational child's tendency to attribute human thoughts, motives, and feelings to non-human objects or organisms. We here use the term *adultomorphic* to describe educators', psychologists', and other adults' tendency to attribute *adult* thoughts, motives, and feelings to nonadult organisms.

An example from our interviews of adolescents should concretize this point. Most health-related researchers—and other reasonably rational adults—think about, measure, and write about initial encounters with cigarettes and alcohol as if they were sharply defined "instances." These instances are then rendered metrically into phrases such as "age of having the *first* alcoholic beverage" or "age of smoking the *first* cigarette." But when you ask teenagers, they are much more likely to talk about their first *cigarette* in terms of their "first smoking *experience*"—"when we got together at Laura's house when her folks were away for the weekend and smoked a lot"—or refer to their first drink in terms of their "first drinking *experience*"—"when we cruised Westheimer all day and night Saturday and drank a lot of Cuervo and Corona." So, researchers—since it serves the precision of their science—think in terms of *instances* and *assume* that teenagers think the same way.

This is a curious form of egocentrism that, in a slightly different form, slowed down progress in developmental psychology for some time—the notion that the child's world is mainly an impoverished and slightly incorrect version of the adult's; that the child's thinking is essentially like that of an incompetent adult. Simply put, we think it is a serious mistake to assume that the subjective world of the teenager is a weaker, somewhat inaccurate, somewhat incompetent version of our own. To the extent that we believe—or behave *as if* we believe—that the adolescent's *umwelt* (vonUexkull, 1957) is nearly isomorphic to our own, we will continue to ask the same questions we have been asking. If we consider seriously the possibility that adolescents inhabit a different reality, we can perhaps make more significant headway into understanding their everyday cognition.

ACKNOWLEDGMENTS

The research reported and the preparation of the manuscript were made possible by a grant from the Texas Affiliate of the American Heart Association (G227) to the senior author through Baylor College of Medicine. We would like to thank the teachers and administrators in both the Cypress-Fairbanks and Victoria Independent School Districts. We thank Thomas Leffler for his valuable assistance throughout the project. We also thank Hayne Reese for his valuable editorial help and substantive suggestions.

REFERENCES

Barker, R. G. (1968). *Ecological psychology.* Stanford, CA: Stanford University Press.
Barker, R. G., & Wright, H. F. (1955). *Midwest and its children.* New York: Harper & Row.
Baumrind, D. (1987). Familial antecedents of adolescent drug use: A developmental perspective. In

C. L. Jones & R. J. Battjes (Eds.), *Etiology of drug abuse: Implications for prevention*. Rockville, MD: U.S. Government Printing Office. (NIDA Research Monograph #56)
Best, J. A., Flay, B. R., Towson, S. M. J., Ryan, K. B., Perry, C. L., Brown, K. S., Kersell, M. W., & d'Avernas, J. R. (1984). Smoking prevention and the concept of risk. *Journal of Applied Social Psychology, 14,* 257–273.
Blos, P. (1941). *The adolescent personality: A study of individual behavior.* New York: Appleton-Century.
Blos, P. (1979). *The adolescent passage: Developmental issues.* New York: International Universities Press.
Blumenthal, A. L. (1970). *The process of cognition.* Englewood Cliffs, NJ: Prentice-Hall.
Blumenthal, A. L. (1975). A reappraisal of Wilhelm Wundt. *American Psychologist, 30,* 1081–1088.
Bretherton, I., Fritz, J., Zahn-Waxler, C., & Ridgeway, D. (1986). Learning to talk about emotions: A functionalist perspective. *Child Development, 57,* 529–548.
Chassin, L., Corty, E., Presson, C., Olshavsky, R., Bensenberg, M., & Sherman, S. (1981). Predicting adolescents' intentions to smoke cigarettes. *Journal of Health and Social Behavior, 22,* 445–455.
Coleman, J. C. (1961). *The adolescent society.* New York: The Free Press.
Coleman, J. C. (1978). Current contradictions in adolescent theory. *Journal of Youth and Adolescence, 4,* 349–358.
Csikszentmihalyi, M., & Larson, R. (1984). *Being adolescent: Conflict and growth in the teenage years.* New York: Basic Books.
Davis, K. (1940). The sociology of parent-youth conflict. *American Sociological Review, 5,* 523–535.
DiClemente, R. J., Zorn, J., & Temoshok, L. (1986). Adolescents and AIDS: A survey of knowledge, attitudes and beliefs about AIDS in San Francisco. *American Journal of Public Health, 76,* 1143–1145.
Douvan, E., & Adelson, J. (1966). *The adolescent experience.* New York: Wiley.
Ekman, P. (1973). *Darwin and facial expression: A century of research in review.* New York: Academic Press.
Elkind, D. (1967). Egocentrism in adolescence. *Child Development, 38,* 1025–1034.
Elkind, D. (1985). Egocentrism redux. *Developmental Review, 5,* 218–226.
Elkind, D., & Bowen, R. (1979). Imaginary audience behavior in children and adolescence. *Developmental Psychology, 15,* 38–44.
Enright, R. D., Lapsley, D. K., & Shukla, D. (1979). Adolescent egocentrism in early and late adolescence. *Adolescence, 14,* 687–695.
Erikson, E. (1951). *Childhood and society.* New York: Norton.
Flavell, J. H. (1963). *The developmental psychology of Jean Piaget.* New York: Van Nostrand.
Furby, L., & Beyth-Marom, R. (1992). Risk taking in adolescence: A decision-making perspective. *Developmental Review, 12,* 1–44.
Grotevant, H. D., & Cooper, C. R. (1985). Patterns of interaction in family relationships and the development of identity exploration in adolescence. *Child Development, 56,* 415–428.
Hall, G. S. (1904). *Adolescence: Its psychology and its relations to anthropology, sociology, sex, crime, religion and education* (2 Vols.). New York: Appleton.
Hartshorne, H., May, M. A., & Maller, J. B. (1929). *Studies in the nature of character.* New York: Macmillan.
Heyneman, S. (1976). Continuing issues in adolescence: A summary of current transition to adult debates. *Journal of Youth and Adolescence, 5,* 309–323.
Hill, J., & Holmbeck, G. (1987). Disagreements about rules in families with seventh grade girls and boys. *Journal of Youth and Adolescence, 16,* 221–246.
Hill, J., & Palmquist, W. (1978). Social cognition and social relations in early adolescence. *International Journal of Behavioral Development, 1,* 1–36.

Izard, C. E. (1971). *The face of emotion.* New York: Appleton-Century-Crofts.

Kandel, D. B., & Logan, J. A. (1984). Patterns of drug use from adolescence to young adulthood: 1. Periods of risk for initiation, continued use and discontinuation. *American Journal of Public Health, 74,* 660–665.

Keating, D. P. (1980). Thinking processes in adolescence. In J. Adelson (Ed.), *Handbook of adolescent psychology* (pp. 211–246). New York: Wiley.

Kett, J. F. (1977). *Rites of passage: Adolescence in America, 1790 to the present.* New York: Basic Books.

Lapsley, D. K. (1990). Continuity and discontinuity in adolescent social cognitive development. In R. Montemayor, G. R. Adams, & T. P. Gullotta (Eds.), *From childhood to adolescence: A transitional period?* (pp. 183–204). Newbury Park, CA: Sage.

Lapsley, D. K., & Murphy, M. N. (1985). Another look at the theoretical assumptions of adolescent egocentrism. *Developmental Review, 5,* 201–217.

Larson, R., & Richards, M. H. (Eds.). (1989). Special Issue: The changing life space of early adolescents. *Journal of Youth and Adolescence, 18*(6).

Lavery, B. S., Siegel, A. W., Cousins, J. H., & Rubovits, D. S. (1993). Adolescent risk-taking: An analysis of problem behaviors in problem children. *Journal of Experimental Child Psychology, 55.*

McCord, W., & McCord, J. (1956). *Psychopathy and delinquency.* New York: Grune & Stratton.

McLaughlin, R., Baer, P., Burnside, M., & Polorny, A. (1985). Psychosocial correlates of alcohol use at two age levels during adolescence. *Journal of Studies on Alcohol, 46,* 212–218.

Napier, A. W., & Whitaker, C. A. (1985). *The family crucible.* New York: Bantam.

Newman, I. M., & Ward, J. M. (1989). The influence of parental attitude and behavior on early adolescent cigarette smoking. *International Journal of the Addictions, 21,* 739–766.

Nolte, A. E., Smith, B. J., & O'Rourke, T. (1983). The relative importance of parental attitudes and behavior upon youth smoking behavior. *Journal of School Health, 53,* 264–271.

O'Connor, B. P., & Nikolic, J. (1990). Identity development and formal operations as sources of adolescent egocentrism. *Journal of Youth and Adolescence, 19,* 149–158.

Offer, D. (1969). *The psychological world of the teen-ager: A study of normal adolescent boys.* New York: Basic Books.

Offer, D., & Offer, J. B. (1975). *From teenage to young manhood.* New York: Basic Books.

Offer, D., Ostrov, E., & Howard, K. I. (1981). *The adolescent: A psychological self-portrait.* New York: Basic Books.

Osgood, C. E., Suci, G. J., & Tannenbaum, P. H. (1957). *The measurement of meaning.* Urbana, IL: University of Illinois Press.

Rogers, P. D., Harris, J., & Jarmuskiwicz, J. (1987). Alcohol and adolescence. *Pediatric Clinics of North America, 34,* 289–303.

Savin-Williams, R. C., & Demo, D. H. (1984). Developmental change and stability in adolescent self-concept. *Developmental Psychology, 20,* 1100–1110.

Schlaadt, R. G., & Shannon, P. T. (1986). *Drugs of choice: Current perspectives on drug use.* Englewood Cliffs, NJ: Prentice-Hall.

Schlosberg, H. (1954). Three dimensions of emotion. *Psychological Review, 61,* 81–88.

Schoggen, P. (1991). Ecological psychology: One approach to development in context. In R. Cohen & A. W. Siegel (Eds.), *Context and development.* Hillsdale, NJ: Lawrence Erlbaum Associates.

Selman, R. (1980). *The growth of interpersonal understanding.* New York: Academic Press.

Siddique, C. M., & D'Arcy, C. (1984). Adolescence, stress and psychological well-being. *Journal of Youth and Adolescence, 13,* 459–473.

Siegel, A. W., Bisanz, J., & Bisanz, G. L. (1983). Developmental analysis: A strategy for the study of psychological change. In D. Kuhn & J. A. Meacham (Eds.), *On the development of developmental psychology. Human Development Monographs.* (pp. 53–80). Basel, Switzerland: Karger.

Siegel, A. W., Cousins, J. H., Rubovits, D. S., Cuccaro, P., Parsons, J. T., & Lavery, B. S. (1993). Affective and cognitive components of adolescent risk taking. Manuscript under review, *Journal of Research on Adolescence.*

Siegel, A. W., & White, S. H. (1982). The child study movement: Early growth and development of the symbolized child. In H. W. Reese (Ed.), *Advances in child development and behavior* (Vol. 17, pp. 233–285). New York: Academic Press.

Steinberg, L. (1989). *Adolescence*. New York: Knopf.

Tabachnick, B. G., & Fidell, L. S. (1989). *Using multivariate statistics* (2nd ed.). New York: Harper Collins.

Timpane, M., Abramowitz, S., Bobrow, S., & Pascal, A. (1976). *Youth policy in transition*. Santa Monica, CA: Rand.

vonUexkull, J. (1957). A stroll through the world of animals and men: A picture book of invisible worlds. In C. H. Schiller (Ed. & Trans.), *Instinctive behavior* (pp. 5–80). New York: International Universities Press.

Wertsch, J. V. (1985). *Vygotsky and the social formation of mind*. Cambridge, MA: Harvard University Press.

White, S. H., & Siegel, A. W. (1976). Cognitive development: The new inquiry. *Young Children, 31,* 425–435.

White, S. H., & Siegel, A. W. (1984). Cognitive development in time and space. In B. Rogoff & J. Lave (Eds.), *Everyday cognition: Its development in social context*. (pp. 238–277). Cambridge, MA: Harvard University Press.

Wolford, C., & Swisher, J. D. (1986). Behavioral intention as an indicator of drug and alcohol use. *Journal of Drug Education, 16,* 305–326.

Youniss, J., & Smollar, J. (1985). *Adolescent relations with mothers, fathers, and friends*. Chicago: University of Chicago Press.

10 Multiple Mechanisms Mediate Individual Differences in Eyewitness Accuracy and Suggestibility

Jonathan W. Schooler
University of Pittsburgh

Elizabeth F. Loftus
University of Washington

In a widely cited article, a simple experiment with 7-year-old students was described (Varendonck, 1911; translated by Hazan, Hazan, & Goodman, 1984). The students were asked to think about a very familiar teacher, Mr. H. The experimenter then asked, "What is the color of Mr. H's beard?" Of 18 students, 16 wrote "black." The other two wrote nothing. In reality, Mr. H. had no beard. Varendonck's (1911) conclusion from this and similar studies was: "a question asked badly can result in erroneous information about a person that children see several times a day" (Hazan et al., 1984, p. 28). A major theme of Varendonck's (1911) work was the idea that when children try to describe the past, their imaginations "play nasty tricks on them"; and that with badly posed questions—whether voluntary or involuntary—we can "obtain answers that stupefy" (Hazan et al., 1984, p. 29).

Why were some of the students misled by the experimenter's question while others were not? To address this question requires that we recognize up front that the recollections of two individuals to the same event often differ. Not only does the general level of accuracy vary, but also the distortions caused by badly posed questions. Such wide variability in performance has been noted since early in this century (Whipple, 1909), and we now take for granted the existence of individual differences in eyewitness testimony (Shapiro & Penrod, 1986). But we probably know less about individual differences in memory for complex past events than we would like. The present chapter reviews some recent efforts to remedy this gap in our knowledge.

Our interest in memory for complex past events must necessarily begin by restricting what we mean by complex past events. Although one might consider a list of words or isolated pictures to be both complex and past, we confine our analyses to studies that involve stimuli that more closely approximate those experienced by real-world "witnesses." In the research reviewed here, subjects experience simulated crimes or accidents or other complex events. In some cases they later get misleading information about the event. Finally, their ability to remember the original event is measured. There is no question that subjects who are faced with this experimental setting vary widely in both accuracy and suggestibility. Some are highly influenced by misleading suggestions whereas others are highly resistant. Is it natural to wonder why? What subgroups, if any, tend to be highly accurate, or highly resistant to suggestion? Conversely, what groups are inaccurate and highly susceptible? Are there any groups that are generally accurate if not confronted with misinformation, but susceptible to its influence when they are?

Questions about individual differences can be about demographic variables, such as age and occupation. Or our concern can be with cognitive or personality variables, such as the extent to which a person has good visual imagery. Researchers have explored both aspects of the individual differences domain in hopes of learning more about the mechanisms that mediate individual differences. In the following discussion we first review a number of demographic variables that have been shown to be associated with eyewitness accuracy and suggestibility. We focus in particular on a recent study that used a science museum as the setting for testing the eyewitness abilities of diverse individuals who visited the museum. Although this demographic review provides valuable information on the relationship between certain demographic variables (such as age and memory performance), it offers only hints at the kinds of cognitive or personality variables that might underlie these differences. In the second section of this chapter, we review the cognitive and personality variables. We also report the findings of an unpublished doctoral dissertation that examined fourteen cognitive and personality variables that might be associated with eyewitness accuracy and suggestibility.

Who Remembers Best?: Demographic Differences

Age is, of course, one demographic variable of great interest to researchers (e.g., Davies, Tarrant, & Flin, 1989). Studies examining the children's eyewitness abilities have had rather mixed results. With respect to overall accuracy, researchers have observed that children spontaneously recall less information than adults (e.g., Dent, 1988; Goodman, Aman, & Hirschman, 1987). However, of the information that children do recall, the proportion that is correct is often equivalent to that of adults (e.g., Goodman et al., 1987; Marin, Holmes, Guth, & Kovac, 1979). With respect to recognition (e.g., answering yes/no questions), some studies have observed that children can be as good as adults (Marin et al., 1979; Saywitz, 1987); whereas, other studies have found that children are less accurate (e.g., Cohen & Harnick, 1980). It is not clear why some studies have found recognition differences and others have not, although some have speculated that it might be due to the materials (e.g., Loftus & Davies, 1984). For example, it may be that some experiments used events that were more interesting to children and consequently observed superior recognition performance; however, no explicit comparison between the materials used in different studies has been done so such conclusions must be considered quite speculative.

There have also been mixed findings with respect to the degree to which children are particularly susceptible to postevent suggestions (Doris, 1991). Inaccurate suggestions may have both immediate and/or delayed effects on individuals' memory reports. Leading questions can induce the inclusion of false facts into one's immediate response to the question. For example, King and Yuille (1987) exposed subjects to a simulated event including an individual who was not wearing a watch and then asked, "On which arm did the man wear his watch?" Such a question can have multiple effects. Some people may simply reject the questions false presupposition and say in this case, "The man was not wearing a watch." Other people may accept the presupposition and say for example "Oh, the watch was on his left arm." We will refer to this initial acceptance reaction to the misinformation as *immediate misinformation acceptance.* A second issue is whether subjects will incorporate these inaccurate presuppositions into their *subsequent* memory reports; we refer to this as *delayed misinformation retrieval.* In theory, it is possible that individuals might go along with the presuppositions of leading questions but would not actually incorporate those suggestions into their later memory reports. In fact, the evidence hints at the possibility that children and adults may differ with respect to their relative propensities for accepting leading questions and retrieving the inaccuracies of those questions in their subsequent reports.

Dale, Loftus, and Rathbun (1978) demonstrated that children as young as age 4 can be induced to accept misinformation implied by subtle linguistic cues. They found that children were more likely to report having seen nonexistent objects when the question was framed, "Did you see *the* . . ." as opposed to "Did you see *a* . . ." With respect to comparisons between children's and adults' immediate misinformation acceptance, a number of researchers have found that children were more susceptible than are adults (e.g., Cohen & Harnick, 1980; Goodman & Reed, 1986; King & Yuille, 1987).

A slightly more complicated picture arises with respect to whether children and adults differ in their inclination to retrieve the incorrect presuppositions of leading questions in their subsequent memory reports. A variety of researchers have failed to find differences in children's and adults' likelihood to retrieve misinformation (e.g., Cohen & Harnick, 1980; Duncan, Whitney, & Kunen, 1982; Goodman & Reed, 1986; Marin et al., 1979; Saywitz, 1987). However, some researchers have found that children are more likely to retrieve misinformation than are adults (e.g., Ceci, Ross, & Toglia, 1987; Loftus, Levidow, & Duensing, 1991). It is not entirely clear why this discrepancy between studies has been observed. Loftus and Davies (1984) speculated that differences in the relative suggestibility of children and adults might have to do with the relative interest of the critical event used in the different studies. An alternative explanation may involve differences in the degree to which studies rely on processes associated with immediate misinformation acceptance versus delayed misinformation retrieval.

It is worth noting that studies have consistently found children to be more likely than adults to immediately accept misinformation but have less consistently found that children are more apt to retrieve misinformation. This difference may in part result because immediate misinformation acceptance requires the subject to immediately acquiesce to the interviewer's assumptions; whereas, in the case of delayed misinformation retrieval, pressures are reduced because the false presupposition is no longer included in the question being asked. Indeed, there is some evidence that children may be particularly apt to immediately accept misinformation because of their unwillingness to challenge the authority of adults (Ceci et al., 1987). Whereas immediate misinformation acceptance involves processes such as acquiescence that may be more likely to be associated with children than adults, delayed misinformation retrieval requires a process that may be more likely to be associated with adults than with children; namely, the ability to recall the previously suggested information. In short, children's greater tendency to immediately accept misinformation may, in some situations, be compensated for by their lesser ability to later recall that misinformation. If this analysis is accurate, then we would expect that children would be reliably more likely to immediately accept misinformation but not necessarily more likely to retrieve it; and this is, in fact, what the literature suggests. Indeed, the two studies that measured both immediate misinformation acceptance and delayed misinformation retrieval (i.e., Cohen & Harnick, 1980; Goodman & Reed, 1986) observed that children were more likely than adults to initially accept the premise of leading questions, but were no more likely than adults to retrieve the misinformation on a subsequent memory test.

The hypothesis that a reduced memory for the misinformation may offset children's greater inclination to accept misinformation suggests the following prediction: With increasing durations between the presentation of misinformation and the memory test, children's likelihood of retrieving misinformation should approach adult levels. This prediction is based on the fact that, with increasing durations between presentation of misinformation and test, individuals become less likely to remember the misinformation (Loftus, Miller, & Burns, 1978). Although no study has directly explored whether age effects are different when the test for misinformation is delayed or not delayed, a number of studies provide indirect evidence. Specifically, many of the studies that *have failed to observe* a difference between children's and adults' likelihood to retrieve misinformation have used relatively long durations between the misinformation suggestion and the subsequent test. For example, Saywitz (1987) included a 5-day interval;

Marin et al. (1979), a 2-week interval; and Cohen and Harnick (1980), a 1-week interval. In contrast, studies that *have observed* children to be more likely to retrieve misinformation have typically included shorter durations. For example, Ceci, Toglia, and Ross (1988) presented the misleading suggestions a day after the presentation of the original event (increasing its salience) and only 2 days before testing subjects. In a study conducted in a science museum (Loftus et al., 1991), which we describe shortly, the misinformation suggestions were presented in the same session as the questions that tested whether subjects would retrieve it. Thus the interval was short, and large age effects were observed. Clearly, further research is needed to explore the precise effects of delaying misinformation and the forgetting that ensues. However, the current analysis suggests that this line of research might well explain previous inconsistencies in children's and adults' reported susceptibility to misinformation.

At the other end of the life span, numerous episodic memory studies have revealed that older people perform more poorly than their younger counterparts at eyewitness memory tasks. For example, Farrimond (1968) presented males between 23 and 79 years, scenes recorded on silent film. Later, an unexpected test of memory for the scenes was given. Performance was best for subjects in their mid-40s and then began to decline, showing noticeable decline after the age of 60 years. The elderly have been shown to perform more poorly than their younger counterparts in free recall of details of the event (List, 1986) and in most categories of event recall (Yarmey & Kent, 1980). Interestingly, in List's (1986) study the elder subjects' performance depended on the material. One event depicted both an older and a younger actress. The older subjects were less accurate than the younger subjects in remembering the details of the younger actress but not the older actress. It may be that this difference occurred because the details of the older actress were more relevant to the older subjects; suggesting that at least some portion of the age difference may reflect differences in perceived relevance of the information.

It would be expected that if the elderly have poorer memories for events, they might demonstrate greater susceptibility to misinformation than younger subjects. Little research has explicitly examined the elderly subjects' relative susceptibility to misinformation. Cohen and Faulkner (1989) recently observed that elderly subjects were, in fact, more likely to retrieve misleading postevent suggestions than were younger subjects. These investigators argued that the decline in resistance to misinformation with age may be due in part to elderly individuals' decreased ability to recall the source of information. Cohen and Faulkner (1989) observed that, compared to younger subjects, elderly subjects had more difficulty recalling the origin of events; that is, whether an event had been performed, imagined, or watched. Further evidence that source attribution becomes more difficult with age is provided by McIntyre and Craik (1987) who observed that older people showed greater inclination to forget the source of facts given to them than did younger people.

Other Demographic Differences

Relatively less research has been devoted to determining the relationship between other demographic differences and eyewitness performance. With respect to gender, there is some evidence that, while men and women have generally comparable eyewitness memory abilities, they do differ with respect to the type of information that they are most apt to remember. For example, Powers, Andriks, and Loftus (1979) observed that women were more accurate and more resistant to suggestions about female oriented details (e.g., clothing) whereas men were more accurate and resistant to suggestions about male oriented details (e.g., type of car). Other research has also supported the conclusion that the two sexes differ in terms of what type of information they remember best (for a review, see Loftus, Banaji, Schooler, & Foster, 1987).

Very little research has examined whether any other demographic variables (such as occupation or education) might affect eyewitness ability. The only occupation that has been given any consideration with respect to this question is law enforcement. Typically, lay people commonly believe that police officers have better memory abilities than civilians (Loftus, 1984; Yarmey & Jones, 1983). However, empirical studies examining this issue have been highly equivocal. In some studies, the police performed similarly to nonpolice (Ainsworth, 1981). In other studies, police made more errors, primarily because they were biased towards "seeing" crimes happen, even when they didn't (Clifford, 1976).

THE MUSEUM STUDY

Although eyewitness ability has been examined with respect to a number of different demographic variables, these studies typically examine only a single variable such as age; and the comparison groups are almost always college students, making comparisons between different groups quite difficult. Moreover, a number of interesting demographic variables such as education and socioeconomic status have been overlooked entirely.

In order to get a more complete picture of the demographic factors that influence eyewitness ability, a recent study by Loftus et al. (1991) examined museum-goers' memories for a taped incident presented at a San Francisco science museum. The experiment was one of the interactive exhibits for visitors to the museum, which means that they provided data for the experiment while simultaneously learning from the exhibit. All visitor-subjects watched a short film clip and later answered a series of questions about it. Moreover, some subjects were exposed to misleading questions; while others were not, so that the impact of misinformation could be assessed. Four demographic questions gathered information about the gender, age, occupation, and educational level of the subject. Statistical analyses then revealed relationships between the demographics and performance (accuracy and suggestibility). In order to give the reader a feel for this novel method of data collection, we briefly review the procedure.

Method. When subjects entered the museum, they watched a short film clip taken from the motion picture "Z," displayed near the museum entrance. The film clip depicted a political rally at which the protagonist walks through a tense, crowded town square. He is threatened by a number of people, and then is rushed by a menacing blue vehicle. In the end, a man at the rear of the vehicle strikes the hero with a night stick, and the man falls to the ground.

After viewing the film clip, subjects wandered at their own pace towards the Memory exhibit, which was at the back of the museum. Once at the exhibit, a large sign enticed subjects to sit down in front of a Macintosh Computer. All subjects were asked a series of 10 questions about the film, all of which could be answered "yes" or "no." For approximately half of the subjects, one early question presented misleading information about the color of a key object. The question was: "Did you see the two thugs in the black shirts who threatened the victim just before the white vehicle tried to run him over?" Actually the vehicle was blue. The remaining subjects answered an early question that did not contain misleading information. For all subjects, a later question asked whether they had seen that the charging vehicle was white. Thus, this study measured delayed retrieval of misinformation. After answering the questions, subjects were debriefed via computer. After the debriefing, demographic information was collected on occupation, education, age, and gender.

Results. From the data, two scores were calculated, one for accuracy and one for suggestibility. The accuracy score was based on five "filler" items; and across all 1989 subjects, performance was about 74% correct. In terms of suggestibility, about a quarter of the subjects claimed that they saw a white, rather than blue, vehicle. They were significantly more likely to erroneously claim they saw a white vehicle when they had received misinformation than when they had not.

Age. The age of subjects varied between 5 and 75 and was systematically related to performance. Table 10.1 shows that accuracy rose as a function of age, up to the 26–35 year-old range, and then began to fall.

In terms of age and suggestibility, the youngest and the oldest age groups were highly suggestible. Put another way, they were significantly more accurate when not misinformed. The 5–10 year-old's, for example, showed a 19% misinformation effect; the over-65 year-old's showed a 32% misinformation effect.

Gender. There was no significant difference between the accuracy of males and females. Interestingly, males, but not females, were significantly misinformed. Given that the critical suggestibility item was a vehicle and males have

TABLE 10.1
Percentage of Ss Giving Misinformation Answer

Age Group	*Control*	*Misinformed*
5-10	18.9 (37)	37.8 (45)
11-12	48.7 (39)	29.6 (44)
13-15	26.1 (46)	22.9 (48)
16-18	36.8 (76)	25.9 (58)
19-25	20.9 (254)	35.0 (306)
26-35	19.3 (264)	30.1 (302)
36-50	29.7 (199)	22.5 (182)
51-65	60.9 (23)	21.7 (23)
Over 65	38.5 (52)	70.0 (30)

Number in parenthesis denotes n for that cell. From Loftus et al. (1991)

TABLE 10.2
Performance on the Five Accuracy Items as a Function of Occupation

Occupation	*Number of Subjects*	*Percent Correct*
Unemployed	60	65.3
Lawyer	80	66.5
Law enforcement	25	68.0
Retired	14	70.0
Student	786	73.6
Homemaker	53	74
Artist/Architect	126	74.1
Scientist	109	74.7
Medical	126	75.1
Business	281	75.4
Teacher	118	75.8
Sales	67	76.4
Trade/Technical	144	76.5

From Loftus et al. (1991)

TABLE 10.3
Percentage of Ss Giving Misinformation Answer

Occupation	*Control*	*Misinformed*
Student	24.9 (394)	28.1 (392)
Artist/Architect	16.7 (48)	53.9 (78)
Teacher	32.7 (52)	34.9 (66)
Lawyer	29.2 (48)	25.0 (32)
Medical	16.7 (60)	31.8 (66)
Business	26.4 (140)	27.7 (141)
Trade/Technical	25.6 (78)	19.7 (66)
Sales	21.4 (14)	35.9 (53)
Scientist	30.8 (52)	35.1 (57)
Homemaker	50.0 (30)	13.0 (23)
Unemployed	28.0 (25)	40.0 (35)

Number in parenthesis denotes n for that cell. From Loftus et al. (1991)

been previously shown to be more interested and more accurate about vehicles (Powers et al., 1979), this finding is somewhat puzzling.

Education. Individuals with higher educational levels showed, on the average, better memory performance than individuals with lower levels. Although significant, the magnitude of this difference was not very great. There was no significant difference in the effects of misinformation between the lowest and highest educational levels. Interestingly though, individuals with higher education levels exhibited a numerically greater effect of misinformation. In short, compared to the rest of the world, people with higher educational levels (i.e., 1–2 years or more of college) tended to be more accurate but comparably suggestible.

Occupation. As can be seen in Table 10.2, Accuracy also varied as a function of occupation. Percent correct ranged from a low of 65.3% for the unemployed to a high of 76.5% for those who classified themselves as "trade/technical." There was no evidence that individuals who described themselves as involved in "law enforcement" had particularly good memories; to the contrary, a test combining lawyers and law enforcement persons showed that they performed less well than the rest of the population. In contrast, relatively good performance was exhibited by teachers, artists, and architects (although they were not significantly better than average).

The relationship between occupation and misinformation provided a rather paradoxical discovery. Specifically, it was observed that a number of the occupations that were associated with relatively good accuracy were surprisingly susceptible to misinformation (See Table 10.3). For example, artist/architects, whose accuracy tended towards being above average, showed the greatest susceptibility to misinformation.

Apparently, while having a bad memory for event details can be associated with greater susceptibility to misinformation (as in the case of the elderly and children), having a good memory (as in the case of people with higher education and artist) does not necessarily prevent one from being susceptible to misinformation.

One possible resolution to the paradox that individuals with both good and bad memories are relatively susceptible to misinformation may lie once again in the distinction between immediate misinformation acceptance and delayed misinformation retrieval. Specifically, children and the elderly may be particularly apt to accept the misleading suggestion because, not remembering the original detail, they will be less likely to notice that the misinformation is discrepant. People with good memories, such as college students and artists and architects, may be relatively less likely to accept the misinformation because they have better memories for the original event. However, if they don't notice that the misinformation is discrepant they may be particularly likely to retrieve the misleading information, due to their generally good memories. This hypothesized proficiency at remembering the misinformation may account for why individuals with good memories are not particularly resistant to retrieving misinformation. As will be seen, the distinction between immediate misinformation acceptance and delayed misinformation retrieval may also help to account for a variety of puzzles regarding the relationship between cognitive variables in eyewitness accuracy and suggestibility.

COGNITIVE AND PERSONALITY DIFFERENCES

There are a number of explanations to call upon in speculating about the impact of various demographic factors on eyewitness memory and suggestibility (e.g., Lindsay & Johnson, 1987). However, with a few exceptions, most of the studies that have looked at demographic differences have not systematically examined the various cognitive and personality variables that have been postulated to mediate these differences. In order to formulate reasonable hypotheses about which individual variables can account for demographic differences in eyewitness memory, it is necessary to develop an understanding of what general cognitive and personality variables are important. Without such knowledge, one cannot be certain that their postulated mechanisms have any explanatory power. For example, it would be fruitless to suggest that increases in working memory capacity are responsible for developmental improvement in eyewitness memory performance if there is, in fact, no relationship between working memory capacity and eyewitness memory. In the following section, we review a number of published studies that have explored which cognitive and personality variables may be important in mediating individual differences in eyewitness memory and suggestibility. We then describe some unpublished research that sheds light on the issue.

General Eyewitness Ability

Surprisingly, few cognitive or personality variables have been found to be predictive of subjects' general ability to remember details of events. Powers et al. (1979) examined the relationship between subjects' scores on the Washington Pre-College Test (WPC) and their memory for details in a 24-slide series depicting a wallet snatching. The WPC test is routinely given to college-bound high school students in Washington State and generally similar to the nationwide Scholastic Aptitude Test (SAT). The WPC is composed of nine subtests including: vocabulary, English usage, spelling, reading comprehension, quantitative skills, applied mathematics, mathematic achievement, spatial ability, and mechanical reasoning. Three additional composite scores, English, verbal, and quantitative, are derived from the subtests. Powers et al. (1979) found no relationship between any of the scores on the WPC and eyewitness performance, thus suggesting that intellectual abilities are not predictive of eyewitness ability.

Other researchers have been equally unsuccessful predicting eyewitness accuracy from personality or cognitive styles measures. For example, Christiaansen, Ochalek, and Sweeney (1984) examined the relationship between eyewitness accuracy and two cognitive style variables: field dependence and locus of control. They hypothesized that field dependent individuals, who are allegedly more attentive to socially relevant information, might be better at remembering the appearance of people; whereas, field independents might be better at remembering objects and actions. Moreover, they thought that subjects with the perception of an internal locus of control might outperform those with an external locus due to their alleged greater need to seek out cues in order to manipulate situations. However, neither locus of control nor field dependence were found to reliably predict eyewitness accuracy. The only relationship found was that field independent subjects were better able to judge whether or not their memory performance was accurate.

Introversion/extroversion has also been hypothesized to influence eyewitness ability. For example, Clifford and Scott (1978) hypothesized that introverts, being more easily aroused, might have more difficulty remembering events, particularly if they are arousing. However, they found no evidence to support this hypothesis.

One variable that has been found to be predictive of general eyewitness performance is general observation ability. Boice, Hanley, Shaughnessy, and Gansler (1982) showed subjects a videotape depicting a robbery and two videotapes depicting situations that would be meaningful to college students: a classroom lecture and a casual request for a date. They found that subjects who were accurate in recalling the details of the purse snatching were also able to recall more details from the other two events. This finding suggests that there may be some general observational ability that helps to mediate eyewitness ability. However, caution must be used in interpreting this finding since subjects were shown all three videotapes on the same day. Thus, it is uncertain whether the consistencies that subjects demonstrated were due to true differences in ability or simply their relative level of attentiveness on that particular day.

Suggestibility

Although researchers have had little success identifying individual differences that are predictive of general eyewitness ability, a number of variables have been found to be predictive of eyewitness suggestibility; that is, the degree to which subjects can be manipulated by the experimenter to incorrectly report the details of an event. Once again in exploring the individual differences underlying susceptibility to misinformation, it will be helpful to discriminate between immediate misinformation acceptance and delayed misinformation retrieval.

Immediate Misinformation Acceptance. Perhaps the most elaborate model of the individual differences that predispose individuals to accept the presuppositions of leading questions has been outlined by Gudjonsson and colleagues (e.g., Gudjonsson, 1983, 1984a, 1984b; Gudjonsson & Clarke, 1986; Singh & Gudjonsson, 1984). Gudjonsson (1983) developed an interrogator suggestibility measure that reflects the degree to which a witness (or in this case a listener to a story) is inclined to be manipulated by leading questions. One method that Gudjonsson has used to induce inaccurate responses is to ask questions that suggest affirmative answers even though such answers are incorrect. For example, asking subjects, "Did the woman's handbag get damaged in the struggle?" when it was not actually damaged. Note in this case the question is not inaccurate per se, as in the case of the standard misinformation questions, rather it is simply leading. Another method used by Gudjonsson induces inaccurate responses by asking questions that do not have correct alternatives; for example, asking subjects, "Did the woman hit one of the assailants with her fist or handbag?" when neither alternative occurred. In this case, if either alternative is answered, then the subject is inaccurate.

Gudjonsson used the term "interrogator suggestibility" to refer to subjects' acceptance of the above leading questions. He observed that individuals' score on the interrogator suggestibility scale was predictive of performance in actual interrogation situations. For example, in criminal trials, subjects who retracted confession statements received higher scores than subjects who did not confess despite incriminating forensic evidence (Gudjonsson, 1984b). Gudjonsson has also found high suggestibility scores, indicating greater susceptibility to misinformation, are associated with a number of personality and intellectual abilities including: low self-esteem (Singh & Gudjonsson, 1984); high neuroticism, willingness to lie in order to appear more socially appropriate, low intellectual ability, poor memory performance (Gudjonsson, 1983), and acquiescence (Gudjonsson, 1986). Based on these and other observations, Gudjonsson and Clark (1986) developed an elaborate model of interrogator suggestibility which assumes that whether individuals are inclined to accept incorrect information depends on their cognitive set and their coping resources. Cognitive set involves expectations about the interrogation: previous experiences, general attitude towards the police, etc. If these expectations lead the witness to trust the interrogator and to believe that a particular response is desired or appropriate, then this will predispose them towards suggestibility. Another important factor is one's coping resources that are determined by memory ability, intellectual functioning, field dependence, and self-concept. These factors may contribute to witnesses' inability to trust their own memories and their consequent predisposition to accept the suggestions of the interrogator.

Delayed Misinformation Retrieval. Although Gudjonsson's approach is intuitively appealing and supported by a number of studies, his findings have not always been consistent with those of studies examining suggestibility as determined by whether subjects reproduce misleading postevent suggestions in their subsequent memory retrievals. For example, Tousignant, Hall, and Loftus (1986) found that subjects who retrieved more misinformation were just as accurate in retrieving correct information as subjects who did not retrieve the misinformation. Schooler and Loftus (1986) suggested that this discrepancy might have resulted from a confound in Gudjonsson's (1983) analysis. Specifically, Gudjonsson's (1983) evidence that poor general memory ability is associated with greater susceptibility to misinformation was based on subjects' memory for the same items on which the misleading suggestions were later directed. Because suggestibility on particular items is likely to be associated with poor memory for those items, Gudjonsson's analysis did not provide an independent measure of general memory ability. In response to Schooler and Loftus' (1986) criticism, Gudjonsson (1987) conducted a new study examining suggestible individuals' memories for items on which they had not been misled. This study found more persuasive evidence that subjects who have high interrogator susceptibility do in fact have generally poorer memories, thus revealing a true inconsistency between the characteristics predictive of whether individuals will initially accept misinformation and whether they will retrieve it as part of their recollection of the event.

Another variable to consider in thinking about immediate misinformation acceptance versus delayed misinformation retrieval is intelligence. At least one study showed a negative correlation between intelligence and subjects' acceptance of the suggestions of an interrogator (Gudjonsson, 1983), while another study showed no relationship between performance on the Washington Pre-College test and subjects' inclination to retrieve previously read misinformation (Powers et al., 1979). A reasonable interpretation of this apparent discrepancy involves the different memory demands that occur when people accept the premises of leading questions versus when they retrieve misinformation in subsequent memory reports. Accepting the premise of a question does not require much memory. In contrast, delayed misinformation retrieval requires recalling the previously encountered misinformation. Thus, whether or not subjects can recall the misinformation at all plays a crucial role in delayed misinformation retrieval but practically no role in immediate misinformation acceptance. If, as Gudjonsson argues (1983), poorer memory and intellectual abilities are associated with a reduced ability to resist the suggestions of an experimenter, then when immediate misinformation acceptance is the dependent measure, such individuals should be more susceptible. However, individuals with poorer intellectual and memory abilities may have more difficulty remembering the presuppositions that they previously agreed with, thus counteracting their greater predisposition to accept the misinformation in the first place. Note that this explanation is analogous to the one we used earlier to explain differences in children's and adults' susceptibility to misinformation.

Although the distinction between immediate misinformation acceptance and delayed misinformation retrieval seems to be useful in accounting for discrepancies in the findings on individual differences and suggestibility, this distinction is not typically made. The vast majority of studies examining individual differences in susceptibility to misinformation have focused on whether or not subjects retrieve previously suggested misinformation. Since the inclusion of retrieved misinformation in one's memory report, requires that it has been initially accepted, these studies do not reveal whether the predicting variables are associated with misinformation acceptance, misinformation retrieval, or both. Nevertheless, a review of these studies does help to shed light on the cognitive and personality variables that predispose an individual to falsely remember inaccurate facts mentioned by postevent sources (i.e., questions or narratives). Moreover, consideration of the distinction between misinformation acceptance and retrieval may help to account for some puzzling findings in a recent dissertation thesis that adds new twists to the picture of the cognitive and personality characteristics that are predictive of eyewitness suggestibility.

Personality Characteristics. Are there personality characteristics that predispose people towards a susceptibility to incorporate misinformation? Unfortunately, answers to this question have been mixed. Ward and Loftus (1985) examined the relationship between subjects' susceptibility to misinformation and their scores on the Myer-Briggs Type indicator (a test based on Jungian personality types). Introverts were found to be more influenced by misinformation than were extroverts. A number of possible explanations were given for this finding. Perhaps introverts have less confidence in their memories than extroverts making them more willing to accept misleading postevent suggestions. Perhaps, introverts are more easily aroused by the original event than extroverts, causing them to form a less stable memory representation that is more susceptible to subsequent information. At present, however, neither speculation should be given much weight since a recent study by Trouve' and Libkuman (1991) found the opposite result; that is, extroverts were more susceptible to misinformation than were introverts. It is unclear what procedural variations may be responsible for these conflicting findings making it difficult to speculate about the bases for differences between introverts' and extroverts' susceptibility to misinformation.

A second Jungian personality variable measured by the Myer-Briggs scale that has been more reliably associated with a susceptibility to misinformation is the intuitive/sensate distinction. Individuals whose test scores indicate they are sensates are inclined to obtain information directly from their senses; whereas, individuals whose test scores indicate they are intuitive allegedly have a preference for obtaining information intuitively. Both Ward and Loftus (1985) and Trouve and Libkuman (1991) found, as one might have predicted, that intuitive individuals, with their lesser reliance on sensory information, were more susceptible to misinformation.

Another personality variable that has been researched is hypnotic suggestibility. It might be expected that individuals who are more prone to being hypnotized and accepting hypnotic suggestion might also be more susceptible to inaccurate suggestions contained in postevent sources. Indeed, a number of researchers have observed that hypnotized subjects are particularly susceptible to misleading postevent information (e.g., Putnam, 1979; Zelig & Beidleman, 1981). However, Sheehan and Tilden (1983) found individuals who indicated a high degree of hypnotic suggestibility were no more likely to be influenced by misinformation either awake or while hypnotized than were subjects with low hypnotic suggestibility scores.

Cognitive Abilities. Researchers have also been interested in the cognitive characteristics that may underlie individuals' susceptibility to misinformation. As mentioned earlier, Powers et al. (1979) found no relationship between subjects' intellectual abilities as measured by the Washington Pre College test and their susceptibility to misinformation. In an unpublished study discussed in Schooler and Loftus (1986), we similarly found no relationship between textual working memory capacity (Daneman & Carpenter, 1980) and susceptibility to misinformation.

Although reading ability per se does not appear to correlate with susceptibility to misinformation, Tousignant et al. (1986) observed that reading speed of postevent information had predictive value. Specifically, they found that subjects who slowly read questions containing misinformation were less likely to be influenced by misinformation than subjects who read the questions quickly. Their interpretation of this finding was that subjects who read more slowly are more likely to catch the discrepancies between the text and the original event and are, consequently, less likely to incorporate the misinformation. Of course, an alternative explanation could be that subjects who read more slowly are less good readers and, consequently, less likely to remember the inaccurate suggestions. However, in a subsequent study, subjects were asked to report whether they had noticed any discrepancies, and subjects who read more slowly were, in fact, more likely to notice the discrepancies than were subjects who read the text quickly. Thus, it seems likely that an inclination to read questions slowly may reduce individuals' susceptibility to misinformation by increasing their ability to catch discrepancies.

A MULTIFACTOR INDIVIDUAL DIFFERENCE STUDY

Although a number of studies have examined the individual differences underlying the misinformation effect, most studies have only examined one or two differences making comparisons between measures difficult. Moreover, a number of possibly important variables have not been considered at all. In order to address these issues, Tousignant (1984) conducted an extensive study examining the relationship between 16 different cognitive and personality variables and susceptibility to misinformation.

Personality Variables. A number of researchers have speculated that an important factor in whether or not subjects incorporate postevent suggestions is whether they are predisposed to behave in ways that they feel the experimenter expects (Gudjonsson, 1983, 1986; McCloskey & Zaragoza, 1985); that is, perhaps subjects adopt the misinformation because they think the experimenter wants them to. To address the role of such demand characteristics, Tousignant (1984) used a number of personality tests that measure different ways in which people may react to the social demands provided by a situation. Specifically, these scales measure different ways in which subjects may respond to the social demands associated with taking a personality test. For example, the Edwards Social Desirability Scale (Edwards & Walker, 1961) measures the tendency of individuals to describe themselves in a socially desirable manner on personality inventories. The Marlow-Crowne Social Desirability Scale (Crowne & Marlowe, 1964) taps an individual's tendency to "lie" or "fake good" on personality inventories by endorsing highly improbable self-descriptive statements. Edwards' acquiescence scale (Edwards & Abbott, 1969) reflects the tendency of an individual to acquiesce or agree with an item of self-description.

In addition to examining the degree to which subjects are susceptible to demand characteristics of personality tests, Tousignant (1984) also used measures that attempt to discern subjects' general level of self-consciousness. It seems reasonable that if individuals are self-conscious about how they appear to the experimenter, they may be more inclined to go along with the experimenter's suggestions. Tousignant (1984) employed three self-consciousness scales developed by Fenigstein, Scheir, and Buss (1975). The private self-consciousness scale measures the degree to which subjects are aware of paying attention to the nature of their internal thoughts. The public self-consciousness scale measures the extent to which subjects are concerned about how they appear to others. Finally, the social anxiety scale reflects the discomfort felt in the presence of others.

Cognitive Variables. Tousignant (1984) examined cognitive variables, many of which had never before been examined in connection with misinformation. The relationship between imagery ability and susceptibility to misinformation was of particular interest because plausible cases can be made to suggest that visual imagery ability is associated with high susceptibility as well as low susceptibility to misinformation. If visual imagery allows people to generate vivid images of suggested information, then they may be more inclined to incorporate that information. Alternatively, if visual imagery allows individuals to vividly recall the original event details, then it may facilitate the identification of discrepancies in the postevent information and thereby reduce its effects. Five measures of imagery ability were used. The Vividness of Visual Imagery Questionnaire (VVIQ) developed by Marks (1972, 1973) measures a person's subjective assessment of the vividness of images that they conjure up. Richardson's (1977) Verbalizer–Visualizer Questionnaire measures the degree to which individuals are inclined to represent and process information visually versus verbally. Two additional imagery measures were included that have also been shown to be associated with hypnotic suggestibility. The Betts Mental Imagery Questionnaire, as adapted by Sheehan and Tilden (1983), is similar to the VVIQ with the exception that it asks subjects to assess, in addition to visual images, the vividness of imagining taste, smell, movement, and feelings. Finally, the Absorption scale, created by Tellegen and Atkinson (1974) measures the degree to which individuals are able to completely concentrate on and become absorbed by their mental images.

A second class of cognitive variables that Tousignant (1984) examined were subjective memory measures. A number of researchers have developed tests that ask subjects to report how effective their memories are in everyday situations. While these measures have proven to have limited validity with respect to predicting everyday memory failures (see Herrmann, 1982, for a review), they have not been applied to eyewitness situations. Thus, it is possible that these measures might predict retention of event details and susceptibility to incorrect information about those details. Three subjective memory measures were used. The Short Inventory of Memory Experiences, developed by Herrmann and Neisser (1978), examines people's general beliefs about their memories and is divided into two subscales: SIME R focuses on remembering and the SIME F on forgetting. The Subjective Memory Questionnaire, created by Bennett-Levy and Powell (1980), asks subjects to make goodness-of-memory judgments for a variety of common types of activities, such as spelling or remembering one's place in a book. Finally, the Cognitive Failure Questionnaire, developed by Broadbent, Cooper, FitzGerald, and Parkes (1982) measures self-reported failures in perception, memory, and motor function.

Method. Tousignant (1984) conducted a fairly standard misinformation study involving 144 undergraduates who participated in three experimental sessions. During the first two sessions, subjects were given the various individual difference scales. In the third session, subjects viewed a slide sequence depicting a purse snatching and then were randomly assigned to either the control or misled conditions. Misled subjects received a question that inaccurately referred to certain items in the slide sequence. Control subjects received similar questions; however, they made no reference to the critical items (items that were incorrectly described in the misled subjects' tests). Finally, all subjects were given a true/false recognition test that asked subjects about both the critical items and filler items that had not been mentioned in the earlier test.

Results. Compared to control subjects, subjects in the misled conditions were substantially less accurate in identifying the critical items on the final yes/no test. Thus a typical misinformation effect was obtained. Of particular interest was the relationship between accuracy on the test and subjects' scores on the various individual differences measures. As can be seen in Table 10.1, few variables predicted general eyewitness accuracy; such as, control subjects' performance on all items and misled subjects' performance on filler items. This finding is consistent with previous findings (reviewed in an earlier section) indicating that personality and cognitive variables have relatively little value when it comes to predicting general eyewitness performance.

By contrast, there were a number of factors that did predict susceptibility to misinformation. For example, the VVIQ was associated with misled subjects' accuracy on the critical items, producing a correlation that approached significance ($p = .06$). This correlation suggests that subjects who have greater visual imagery abilities were more inclined to accept misinformation. Paradoxically, subjects who indicated a preference for visual compared to verbal representation and processing on the VVQ were *less* susceptible to misinformation. The reverse relationship between susceptibility to misinformation and performance on the VVIQ and VVQ argues for different processes underlying the two measures.

TABLE 10.4
The Correlation Between Various Individual Difference Measures and the Proportion of Items Recognized by Control and Misinformed Subjects

	Recognition of Critical Items		*Recognition of Filler Items*	
Predictor[1]	*Control*	*Misinformed*	*Control*	*Misinformed*
Sime part F (forgetting)	-.03	-.35*	-.06	.02
Sime part R (remembering)	-.18	-.22	-.08	-.16
SMQ (everyday memory)	.04	-.42*	.18	.11
CFQ (memory lapses)	-.23	.24*	-.05	-.04
VVIQ (visualization ability)	-.10	.22	-.16	.15
VVQ (preference for visual information)	.02	.23*	-.06	-.05
E sd scale (social desirability)	.25*	-.03	-.01	.12
R scale (acquiesence)[2]	.07	.35*	-.08	.21
MC sd scale (lying)	.19	-.01	.07	.03
SCS1 (internal self-consciousness)	-.01	.01	-.09	-.04
SCS2 (external self-consciousness)	.01	-.16	-.05	-.08
SCS3 (social anxiety)	-.31*	.04	-.26*	-.01
BMIQ (vividness of imagination)[2]	.13	.16	-.12	.09
Absorption (total retention)	-.05	-.09	-.10	-.06

From Tousignant (1984).
*$p < .05$.
[1]unless indicated otherwise, all scales are keyed so that higher scores indicate a greater degree of the trait indicated in parentheses.
[2]These scales are keyed so that *lower* scores indicate a greater degree of the trait indicated in the parentheses.

This conclusion is further supported by the lack of a relationship between the VVIQ and the VVQ (see Table 10.5). How shall we explain this apparent puzzle? Perhaps, individuals with high visual imagery are capable of generating vivid images of the suggested items and, consequently, are more influenced by misinformation, explaining the near significant correlations between the VVIQ and susceptibility to misinformation. By contrast, individuals who prefer to represent information verbally are susceptible to misinformation because the misinformation is presented verbally while the original event was visual, thus explaining the correlations between the VVQ and susceptibility to misinformation. Whatever the ultimate explanation, it appears clear that quite different mechanisms likely mediate these two relationships.

The third notable predictor of the effects of misinformation is performance on the acquiescence scale. Recall that this scale measures subjects' willingness to agree with an item of self description even if that item is probably untrue. Thus, this measure reflects the degree to which a person is a "yes person." The interpretation of the positive correlation of acquiescence to misinformation is quite straightforward: individuals who are inclined to agree with things they do not really believe will be more likely to accept suggestions that they saw things that they don't really recall seeing. Thus, the correlation between this measure and suggestibility indicates that some portion of the effect of misinformation can be attributed to subjects' susceptibility to demand characteristics.

The final and perhaps most interesting predictors of susceptibility to misinformation are the subjective memory measures. Surprisingly, the relationship was opposite to what might be expected. Subjects who reported that they had good memories were actually *more* likely to be influenced by the misinformation than subjects who indicated their memories were bad. This relationship was statistically significant for three of the four subjective measures and approached significance (.06) for the remaining measure (SIME part R). Interpretation of this counter-intuitive finding is completely post-hoc. Tousignant (1984) suggested that it might occur because subjects who score high on the subjective memory reports desire to be perceived as good rememberers. He speculated that such subjects might respond to a suggested misinformation item as follows: "Of course I saw that, what do you think I am, an unperceptive observer?" As evidence for this self-presentation interpretation, he cites the high correlation between social desirability measures and subjective memory measures. The problem with this interpretation can be seen in Table 10.2. Specifically, the social desirability measure, that subjects' need to appear competent to the experimenter, was itself not correlated with susceptibility to misinformation. Moreover, acquiescence, the one measure of subjects' sensitivity to demand characteristics that correlated with misinformation susceptibility, did not correlate with the memory measures. Thus, a self-presentation explanation of the predictive value of the memory measures does not seem to be a coherent story.

An alternative reason why people who believe they have good memories are

TABLE 10.5

Self-Report Measures Correlation Matrix[1]

	SF	*SR*	*SMQ*	*CFQ*	*VVIQ*	*VVQ*	*Esd*	*R*	*MC*	*SCS1*	*SCS2*	*SCS3*	*BMIQ*	*ABS*
Sime part F (forgetting)														
Sime part R (remembering)	-.14													
SMQ (everyday memory)	-.45	.37												
CFQ (memory lapses)	.77	-.17	-.45											
VVIQ open (visualization ability)	.22	-.29	-.35	.24										
VVQ (preference for visual information)	.22	-.02	-.16	.28	.11									
E sd scale (social desirability)	-.49	.04	.21	-.60	-.05	-.18								
R scale (acquience)[2]	.17	-.06	-.18	.08	.20	.05	.10							
MC sd scale (lying)	-.37	.20	.39	-.39	-.20	-.04	.40	.14						
SCS1 (internal self-consciousness)	.17	.11	.15	.17	-.16	.03	-.33	-.18	-.05					
SCS2 (external self-consciousness)	.12	.04	.12	.17	-.14	.15	-.26	-.19	-.07	.50				
SCS3 (social anxiety)	.35	-.18	-.33	.45	.04	.28	-.55	.06	-.33	.01	.09			
BMIQ (vividness of imagination)[2]	.33	-.22	-.35	.20	.56	.10	.03	.31	-.06	-.26	-.34	-.01		
Absorption (total retention)	.12	.06	.15	.18	-.24	-.15	-.35	-.39	-.06	.55	.26	.03	.29	

From Tousignant (1984).

[1]Unless indicated otherwise, all scales are keyed so that higher scores indicate a greater degree of the trait indicated in parentheses.

[2]These scales are keyed so that *lower* scores indicate a greater degree of the trait indicated in the parentheses.

more influenced by misinformation may once again result from the memory demands associated with the delayed retrieval of misinformation. Recall that in Tousignant's (1984) study, the misinformation was presented in the context of a memory test. Consider the question: What type of people would be most attentive and interested in a memory test? The answer seems straightforward: those people who think they will do well on it. What type of people think they will do well on a memory test? People who have indicated they think they have good memories; people who have scored highly on the subjective memory tests. In short, people who believe they have good memories may pay closer attention to the memory test in which the misinformation is imbedded and may subsequently be more apt to remember and incorporate this information. This explanation seems quite plausible since these individuals are not particularly likely to remember the original details (as indicated by the lack of a relationship between memory report measures and performance of control subjects). There is also a considerable body of evidence indicating that although subjects who believe they have good memories do not actually possess such memories, they nevertheless behave as if they did (for a review see Herrmann, 1990). For example, people who think they are good at trivia questions, although actually no better than average, are still more willing to wage money in trivia games (Sehulster, 1981). Thus, it seems likely that subjects who believed they had good memories were more apt to relish the opportunity to show off their memory skills and, consequently, paid more attention to the test in which the misinformation was embedded.

The foregoing explanation rests on the assumption that the increased suggestibility of people who believe they have good memories results because these people are more apt to remember the postevent information. Thus, one prediction of this interpretation is that if subjects receive misinformation that does not need to be retained in order for it to exert its influence, as in the case of immediate misinformation acceptance, then the correlation between suggestibility and subjective memory ability should decrease. Further research might profitably compare the degree to which subjective memory reports are predictive of immediate misinformation acceptance and delayed misinformation retrieval.

SUMMARY AND CONCLUSIONS

The pattern of individual differences associated with eyewitness memory and suggestibility are quite complex. With respect to demographics, it is frequently, although not always, observed that the young (less than 10) and the elderly (over 65) tend to be less accurate than people between those age brackets. However, many factors complicate this picture. For example, if young children are given recognition tests and if the material is familiar, their memories may be as good as that of adults. Other demographic differences also appear to influence eyewitness accuracy and performance. Generally speaking, people tend to be better able to remember that type of information with which they are most familiar. For example, older and younger individuals differ with respect to the type of information that they find most relevant and, consequently, are most apt to remember.

Demographic differences in susceptibility to misinformation also reveal some complex differences. These differences suggest a distinction between *immediate misinformation acceptance,* which corresponds to whether or not an individual immediately accepts the inaccurate presupposition of a misleading suggestion, and *delayed misinformation retrieval,* which involves incorporating a misleading suggestion into a subsequent memory report. A number of studies have observed that, compared to adults, children are reliably more likely to immediately accept misinformation when given leading questions. However, studies have differed with respect to children's and adults' relative inclination to retrieve previously suggested misinformation. Some studies have found that children are more apt to retrieve misinformation than are adults; whereas, other studies have not observed this difference. With respect to the elderly, a few studies now indicate that they are relatively more likely to retrieve previously received misinformation.

Results of a recent study by Loftus et al. (1991) revealed some particularly intriguing differences in the likelihood of retrieving misinformation. First, as is often though not always seen, young and elderly subjects were particularly susceptible to misinformation. Because these subjects also had relatively poor memories, the following explanation of their particular susceptibility to misinformation seems warranted; the young and elderly subjects tended not to remember the original detail and consequently not to notice that the postevent information was discrepant. They were therefore more apt to accept the misinformation. This discrepancy detection account, fails however in explaining the relative susceptibility of other demographic groups. Specifically, the susceptibility to misinformation of individuals who have particularly good memories is in some cases greater (e.g., artists and architects) and in other cases equal (e.g., college students) to that of individuals with less good memories. This latter finding produces somewhat of a puzzle. If poor memories lead to a susceptibility to misinformation, then why don't particularly good memories reduce this susceptibility?

An understanding of the mechanisms underlying the above demographic differences requires knowledge of the general cognitive and personality factors that may mediate a susceptibility to misinformation. Here, the literature also reveals considerable complexity. With respect to overall eyewitness accuracy, the only factor that has been shown to have predictive value is general observational skill; and little is currently known about this ability. Greater success has been found in predicting susceptibility to experimenter suggestions. Once again, it appears helpful to distinguish between immediate misinformation acceptance and delayed misinformation retrieval. For example, research by Gudjonsson and colleagues indicates that intelligence and general memory ability are negatively correlated with individuals' propensity to accept misinformation. However, research by Loftus and colleagues suggests that intelligence and memory ability are not correlated with the likelihood that they will later retrieve misinformation. Although memory and intellectual abilities have not been shown to be predictive of an individuals propensity for retrieving misinformation, a variety of other factors have been shown to be positively correlated with delayed misinformation retrieval, including: a reliance on intuitions versus sensation, reading postevent information quickly, good visual abilities, a preference for representing information verbally, the tendency to acquiesce, and perhaps most surprisingly, the belief that one has a *good* memory.

Some of these individual difference variables may help to explain certain demographic differences. For example, the tendency of people with good visual abilities to incorporate misinformation may help to explain why architects and artists appear particularly susceptible to misinformation. The relationship between acquiescence and suggestibility may be one reason why children are particularly susceptible to misinformation when the individual offering the misinformation is an adult rather than a child.

Our consideration of the cognitive variables underlying eyewitness suggestibility led us to the view that distinct mechanisms may mediate individuals' susceptibility to immediate misinformation acceptance and delayed misinformation retrieval. Immediately accepting misinformation, theoretically, requires relatively little memory ability; in fact, poorer memories should make it easier to accept misinformation because individuals will be less likely to detect discrepancies between the misinformation and the original event. This relationship between poor memory ability and immediate misinformation acceptance has been consistently demonstrated both by demographic studies (e.g., comparison of adults and children) and cognitive studies (e.g., Gudjonsson, 1984a). In contrast, delayed misinformation retrieval includes a memory component; that is, individuals must remember the misinformation in order to retrieve it. This may explain why the relationship between memory ability and delayed misinformation retrieval is somewhat more complex. For example, the inconsistent findings regarding age differences in delayed misinformation retrieval may result from differences in the interval between presenting the misinformation and test. Generally speaking, studies with longer intervals have observed less difference between adults and children, perhaps because with longer delays, children are more likely to forget the misinformation, thereby counteracting any greater tendency they may have to accept misinformation in the first place. Similarly, unlike immediate misinformation acceptance, no relationship has been found between intellectual or memory abilities and delayed misinformation retrieval. This may also occur because individuals with good intellectual or memory abilities may be particularly likely to remember, and thus be influenced by, misinformation, thereby counteracting any advantage that they might otherwise have.

The distinct mechanisms underlying immediate misinformation acceptance and delayed misinformation retrieval may also account for why individuals with poorer memory abilities, such as the very young and the elderly, are particularly susceptible to misinformation as are some individuals with relatively good memory abilities, such as artists and architects. People with poor memory abilities may be susceptible to misinformation because they are inclined to forget the original details, and people with good memories are susceptible because they are inclined to remember the postevent information. Finally, the memory demands, hypothesized to underlie delayed misinformation retrieval, may account for why individuals who believe they have good memories are particularly *likely* to retrieve misinformation. Specifically, compared to people with little faith in their memories, people who believe they have good memories may be particularly inclined to show off their memory skills; consequently they may pay closer attention to the memory test questions and thereby become more likely to later retrieve the misinformation embedded in those questions.

The distinction between immediate misinformation acceptance and delayed misinformation retrieval has not been explicitly made before; consequently, the analysis provided here is post-hoc. Nevertheless, this distinction may help to resolve a number of puzzles regarding individual differences in susceptibility to misinformation. If future research demonstrates that the distinction between immediate misinformation acceptance and delayed misinformation retrieval is a useful one, then it will be testament to the theoretical gains that can be made by considering individual differences (Underwood, 1975).

ACKNOWLEDGMENTS

The writing of this chapter was facilitated by two grants from the National Institute of Mental Health, one to Jonathan Schooler at the University of Pittsburgh and one to Elizabeth Loftus at the University of Washington. We thank Douglas Herrmann for his helpful suggestions.

REFERENCES

Ainsworth, P. B. (1981). Incident perception by British police officers. *Law and Human Behavior, 5,* 231–236.

Bennett-Levy, J., & Powell, G. E. (1980). The Subjective Memory Questionnaire (SMQ). An investigation into the self-reporting of 'real-life' memory skills. *British Journal of Social and Clinical Psychology, 19,* 177–188.

Boice, R., Hanley, C. P., Shaughnessy, P., & Gansler, D. (1982). Eyewitness accuracy: A general observational skill? *Bulletin of the Psychonomic Society, 204,* 193–195.

Broadbent, D. E., Cooper, P. F., FitzGerald, P., & Parkes, K. R. (1982). The Cognitive Failures Questionnaire (CFQ) and its correlates. *British Journal of Clinical Psychology, 21,* 1–16.

Ceci, S. J., Ross, D. F., & Toglia, M. P. (1987). Age differences in suggestibility: Narrowing the uncertainties. In S. J. Ceci, M. P. Toglia, & D. F. Ross (Eds.), *Children's memory* (pp. 79–91). New York: Springer-Verlag.

Ceci, S. J., Toglia, M., & Ross, D. F. (1988). On remembering . . . more or less: A trace strength interpretation of developmental differences in suggestibility. *Journal of Experimental Psychology; General, 117,* 201–203.

Christiaansen, R. E., Ochalek, K., & Sweeney, J. D. (1984). Individual differences in eyewitness memory and confidence judgments. *Journal of General Psychology,* in press.

Clifford, B. R. (1976). Police as eyewitnesses. *New Society, 36,* 176–177.

Clifford, B. R., & Scott, J. (1978). Individual and situational factors in eyewitness testimony. *Journal of Applied Psychology, 633,* 352–359.

Cohen, G., & Faulkner, D. (1989). Age differences in source forgetting: Effects on reality monitoring and on eyewitness testimony. *Psychology and Aging, 4,* 10–17.

Cohen, R. L., & Harnick, M. A. (1980). The susceptibility of child witnesses to suggestion: An empirical study. *Law and Human Behavior, 4,* 201–210.

Crowne, D. P., & Marlowe, D. (1964). *The approval motive: Studies in evaluative dependence.* New York: Wiley.

Dale, P. S., Loftus, E. F., & Rathbun, L. (1978). The influence of the form of the question on the eyewitness testimony of preschool children. *Journal of Psycholinguistic Research, 7,* 269–277.

Daneman, M., & Carpenter, P. A. (1980). Individual differences in working memory and reading. *Journal of Verbal Learning and Verbal Behaviour, 19,* 450–466.

Davies, G. M., Tarrant, A., & Flin, R. (1989). Close encounters of a witness kind: children's memory for a simulated health inspection. *British Journal of Psychology, 80,* 415–429.

Dent, H. R. (1988). Children's eyewitness evidence: A brief review. In M. M. Gruneberg, P. E. Morris, & R. N. Sykes (Eds.), *Practical aspects of memory: Current research and issues* (Vol. 1, pp. 101–106). Chichester, England: Wiley.

Doris, J. (Ed.). (1991). *The suggestibility of children's memory: Implications for eyewitness testimony)* Washington D.C.: American Psychological Association.

Duncan, E. M., Whitney, P., & Kunen, S. (1982). Integration of visual and verbal information in children's memories. *Child Development, 53,* 1215–1223.

Edwards, A. L., & Abbott, R. D. (1969). Further evidence regarding the R scale of the MMPI as a measure of acquiescence. *Psychological Reports, 24,* 903–906.

Edwards, A. L., & Walker, J. N. (1961). A short form of the MMPI: The SD scale. *Psychological Reports, 8,* 485–486.

Farrimond, T. (1968). Retention and recall: Incidental learning of visual and auditory material. *Journal of Genetic Psychology, 113,* 155–165.

Fenigstein, A., Scheir, M. F., & Buss, A. H. (1975). Public and private self-consciousness: Assessment and theory. *Journal of Consulting and Clinical Psychology, 434,* 522–527.

Goodman, G., Aman, C., & Hirschman, J. (1987). Child sexual and physical abuse: Children's testimony. In S. J. Ceci, M. P. Toglia, & D. F. Ross (Eds.), *Children's eyewitness memory* (pp. 1–20). New York: Springer-Verlag.

Goodman, G. S., & Reed, R. S. (1986). Age differences in eyewitness testimony. *Law and Human Behavior, 10,* 317–332.

Gudjonsson, G. H. (1983). Suggestibility, intelligence, memory recall and personality: An experimental study. *British Journal of Psychiatry, 142,* 35–37.

Gudjonsson, G. H. (1984a). A new scale of interrogative suggestibility. *Personality and Individual Differences, 5,* 303–314.

Gudjonsson, G. H. (1984b). Interrogative suggestibility: comparison between "False confessions" and "deniers" in criminal trials. *Medicine, Science and the Law, 24,* 56–60.

Gudjonsson, G. H. (1986). The relationship between interrogative suggestibility and acquiescence: Empirical findings and theoretical implications. *Personality and Individual Differences, 7,* 195–199.

Gudjonsson, G. H. (1987). The relationship between memory and suggestibility. *Social Behavior, 2,* 29–33.

Gudjonsson, G. H., & Clark, N. K. (1986). A theoretical model of interrogative suggestibility. *Social Behaviour, 1,* 83–104.

Hazan, C., Hazan, R., & Goodman, G. S. (1984). Translation of Varendonck, J. The testimony of children in a famous trial. *Journal of Social issues, 40,* 26–31.

Herrmann, D. J. (1982). Know thy memory: The use of questionnaires to assess and study memory. *Psychological Bulletin, 92,* 424–452.

Herrmann, D. J. (1990). Self perceptions of memory performances. In W. K. Schaie, J. Rodin, & C. Schooler (Eds.), *Self directedness and efficacy: Causes and effects throughout the life course.* Hillsdale, NJ: Lawrence Erlbaum Associates.

Herrmann, D. J., & Neisser, U. (1978). An inventory of everyday memory experiences. In M. M. Gruneberg, P. E. Morris, & R. N. Sykes (Eds.), *Practical aspects of memory.* New York: Academic Press.

King, M. A., & Yuille, J. C. (1987). Suggestibility and the child witness. In S. J. Ceci, M. P. Toglia, & D. F. Ross (Eds.), *Children's memory* (pp. 24–33). New York: Springer-Verlag.

Lindsay, S. D., & Johnson, M. K. (1987). Reality monitoring and children's ability to discriminate among memories from different sources. In S. J. Ceci, M. P. Toglia, & D. F. Ross (Eds.), *Children's memory* (pp. 24–33). New York: Springer-Verlag.

List, J. (1986). Age and schematic differences in the reliability of eyewitness testimony. *Developmental Psychology, 22,* 50–57.

Loftus, E. F. (1984). Eyewitness: Essential but unreliable. *Psychology Today,* Feb., 22–26.

Loftus, E. F., & Davies, G. M. (1984). Distortions in the memory of children. *Journal of Social Issues, 40,* 51–67.

Loftus, E. F., Banaji, M. R., Schooler, J. W., & Foster, R. A. (1987). Who remembers what? Gender differences in memory. *Michigan Quarterly Review, XXVI,* 64–85.

Loftus, E. F., Levidow, B., & Duensing, S. (1991). Who remembers best? Individual differences in memory for events that occurred in a science museum. *Applied Cognitive Psychology, 6,* 93–107.

Loftus, E. F., Miller, D. G., & Burns, H. J. (1978). Semantic integration of verbal information into visual memory. *Journal of Experimental Psychology: Human Learning and Memory, 4,* 19–31.

Marin, B. V., Holmes, D. L., Guth, M., & Kovac, P. (1979). The potential of children as eyewitnesses. *Law and Human Behavior, 3,* 295–305.

Marks, D. F. (1972). Individual differences in the vividness of visual imagery and their effect on function. In P. W. Sheehan (Ed.), *The function and nature of imagery* (pp. 83–106). New York: Academic Press.

Marks, D. F. (1973). Visual imagery differences in the recall of pictures. *British Journal of Psychology, 64,* 17–24.

McCloskey, M., & Zaragoza, M. (1985). Misleading postevent information and memory for events: Arguments and evidence against memory impairment hypotheses. *Journal of Experimental Psychology: General, 114,* 1–16.

McIntyre, J. S., & Craik, F. I. M. (1987). Age Differences in memory for item and source information. *Canadian Journal of Psychology, 41,* 175–192.

Powers, P. A., Andriks, J. L., & Loftus, E. F. (1979). The eyewitness accounts of females and males. *Journal of Applied Psychology, 64,* 339–347.

Putnam, W. H. (1979). Hypnosis and distortions in eyewitness memory. *International Journal of Clinical and Experimental Hypnosis, 27,* 437–448.

Richardson, A. (1977). Verbalizer-visualizer: A cognitive style dimension. *Journal of Mental Imagery, 1,* 109–126.

Saywitz, K. J. (1987). Children's testimony: Age-related patterns of memory errors. In S. J. Ceci, M. P. Toglia, & D. F. Ross (Eds.), *Children's memory* (pp. 36–50). New York: Springer-Verlag.

Schooler, J. W., & Loftus, E. F. (1986). Individual differences and experimentation: Complementary approaches to interrogative suggestibility. *Social Behaviour, 1,* 105–112.

Sehulster, J. R. (1981). Structure and pragmatics of a self-theory of memory. *Memory & Cognition, 9,* 263–276.

Shapiro, P. N., & Penrod, S. (1986). Meta-analysis of facial identification studies. *Psychological Bulletin, 100,* 139–156.

Sheehan, P. W., & Tilden, J. (1983). Effects of suggestibility and hypnosis on accurate and distorted retrieval from memory. *Journal of Experimental Psychology: Learning Memory, and Cognition, 92,* 283–293.

Singh, K. K., & Gudjonsson, G. H. (1984). Interrogative suggestibility, delayed memory and self-concept. *Personality and Individual Differences, 5,* 203–209.

Tellegen, A., & Atkinson, B. (1974). Openness to absorbing the self-altering experiences ("Absorption"), A trait related to hypnotic susceptibility. *Journal of Abnormal Psychology, 3,* 268–277.

Tousignant, J. P. (1984). *Individual differences in response bias and recall: A characterization of the effects of misleading post-event information.* Unpublished doctoral dissertation, University of Washington.

Tousignant, J. P., Hall, D., & Loftus, E. F. (1986). Discrepancy detection and vulnerability to misleading post-event information. *Memory & Cognition, 14,* 329–338.

Trouve', R. J., & Libkuman, T. M. (1991). *Eyewitness performance of personality types as a function of induced arousal.* Unpublished manuscript, Central Michigan University.

Underwood, B. J. (1975). Individual differences as a crucible in theory construction. *American Psychologist, 30,* 128–134.

Varendonck, J. (1911). Les Temoignages d'enfants dans un proces retentissant. *Archives de Psychologie, 11,* 129–171.

Ward, R. A., & Loftus, E. F. (1985). Eyewitness performance in different psychological types. *The Journal of General Psychology, 112,* 191–200.

Whipple, G. M. (1909). The observer as reporter: A survey of the "psychology of testimony." *Psychological Bulletin, 6,* 153–170.

Yarmey, A. D., & Jones, H. P. T. (1983). Is the psychology of eyewitness identification a matter of common sense? In S. M. A. Lloyd-Bostock & B. R. Clifford (Eds.), *Evaluating witness evidence* (pp. 13–40). Chichester, England: Wiley.

Yarmey, A. D., & Kent, J. (1980). Eyewitness identification by elderly and young adults. *Law and Human Behavior, 4,* 339–371.

Zelig, M., & Beidleman, W. B. (1981). The investigative use of hypnosis: A word of caution. *International Journal of Clinical and Experimental Hypnosis, 29,* 401–412.

11 Practical Intelligence: The Nature and Role of Tacit Knowledge in Work and at School

Robert J. Sternberg
Yale University

Richard K. Wagner
Florida State University

Lynn Okagaki
Yale University

Practical (or everyday) intelligence seems to be a different kettle of fish from academic intelligence. There are any number of ways in which we see this difference in our everyday lives. We see people who succeed in school and who fail in work, or who fail in school but who succeed in work. We meet people with high intelligence-test scores who seem brain-damaged in their social interactions. And we meet people with low test scores who can get along effectively with practically anyone. The research literature confirms our everyday impressions. There just doesn't seem to be much relation between people's academic and practical skills (see Sternberg & Wagner, 1986). For example, Lave, Murtaugh, and de la Rocha (1984) have found that women who can compute effectively in a supermarket price-comparison situation may not be able to compute effectively in a paper-and-pencil arithmetic test of isomorphic operations. Ceci and Liker (1986) have found that men with low-average IQs can show considerable cognitive complexity in their predictions of winners at the race track. Scribner (1986) has shown that men working in a milk-processing plant, probably not men with stunningly high IQs, can be quite innovative in speeding up their work. And at an operational level, we find that whereas conventional intelligence tests predict school performance at a correlational level that is typically in the .4 to .6 range, prediction of job performance is more typically at about the .2 level (Wigdor & Garner, 1982).

There may be any number of reasons for the difference between academic and practical intelligence, but we would suggest that a major source of this difference is the sheer disparity in the nature of the kinds of problems one faces in academic versus practical situations. Academic problems tend to be (a) formulated by others, (b) well-defined, (c) complete in the information they provide, (d) characterized by having only one correct answer, (e) characterized by having only one method of obtaining the correct answer, (f) disembedded from ordinary experience, and (g) of little or no intrinsic interest. Practical problems, in contrast to academic problems, tend to be characterized by (a) the key roles of problem recognition and definition, (b) their ill-defined nature, (c) substantial information seeking, (d) multiple "correct" solutions, (e) multiple methods of obtaining solutions, (f) the availability of relevant prior experience, and (g) often highly motivating and emotionally involving contingencies. Given the differences in the nature of academic and practical problems, it is no surprise that people who are adept at solving one kind of problem may well not be adept at solving problems of the other kind. We therefore might want to seek some construct or set of constructs that would help us frame the difference or differences between the skills needed to solve problems of the two kinds.

This chapter is divided into four main parts. First, we describe the construct of tacit knowledge that motivates all of the work described in the chapter. Second, we describe a series of experiments with adults that illustrates the nature, use, and acquisition of tacit knowledge in a variety of careers, including college professors, business executives, and sales people. Then, we describe experiments done with children that show that tacit knowledge can be taught. Finally, we sum up the main points of our exposition.

THE CONSTRUCT OF TACIT KNOWLEDGE

In academic problems, formal knowledge plays a key role. Any number of studies have shown that expertise in solving academic kinds of problems is heavily dependent on the availability and accessibility of formal knowledge (see, e.g., Chi, Glaser, & Farr, 1988). Formal knowledge seems much less relevant to practical problem solving. Formal knowledge will not tell one, for example, what kinds of things one can and cannot say to a supervisor or a teacher, or how one can best budget one's time to get all of one's job-related tasks done, or how one can avoid procrastinating so as to get these tasks done. In nonacademic tasks, the key kind of knowledge appears to be informal, or what we call *tacit* knowledge (Wagner & Sternberg, 1985, 1986; see also Polanyi, 1976). Tacit knowledge is practical know-how that usually is not directly taught or even openly expressed or stated. It is the kind of knowledge that one picks up on a job or in everyday kinds of situations, rather than through formal instruction. For example, knowing how to convince others of the worth of your idea or product is not a kind of knowledge that is likely directly to be taught, but rather the kind of knowledge one is likely to pick up through experience.

We distinguish among three kinds of tacit knowledge: tacit knowledge about managing oneself, about managing others, and about managing tasks. Tacit knowledge about *managing oneself* refers to knowledge about self-motivational and self-organizational aspects of work-related performance. Tacit knowledge about *managing others* refers to knowledge useful in work-related interactions with one's subordinates, superiors, and peers. Finally, tacit knowledge about *managing tasks* refers to knowledge about how to do specific work-related tasks well. We also distinguish between two orientations of tacit knowledge, depending upon the time frame within which the tacit knowledge would be used. The focus of *local* tacit knowledge is the situation at hand, for example, how to organize the tasks one faces on a given day. The focus of *global* tacit knowledge is on how the situation at hand fits into the bigger picture, for example, how to get a promotion. Because the three kinds of tacit knowledge are orthogonal to the two orientations of tacit knowledge, it is possible to cross them, yielding six (3 × 2) categories in all.

Although our focus in this chapter is on tacit knowledge—its nature, use, and acquisition—we emphasize that the construct of tacit knowledge fits within a more general theoretical framework for understanding intelligence, namely, the triarchic theory of human intelligence (Sternberg, 1985).

The triarchic theory posits that intelligence can be understood in terms of the application of components of information processing to varying levels of experience, which in turn can serve three functions in real-world contexts: adaptation to existing environments, selection of new environments, and shaping of existing environments to turn them into new environments. Tacit knowledge is used in adaptation to environments, but also in deciding when an environment is unsatisfactory and a new one needs to be sought out (environmental selection) or when the present environment can be shaped into a more nearly optimal one (shaping of the environment). Thus, tacit knowledge is most relevant to the contextual or practical subtheory of the triarchic account of intelligence: It is the knowledge base that enables us to face the everyday world.

How does one measure tacit knowledge? Wagner and Sternberg (1985) devised a method of presenting scenarios to individuals that depicted the kinds of problems faced by people in a given life pursuit. Examinees make judgments about these scenarios that require them to have and exploit tacit knowledge. Note that our measures are not simply measures of whether or not individuals have tacit knowledge, but measures of whether they have and then can use it. Table 11.1 shows examples of scenarios measuring tacit knowledge for managing oneself, managing others, and managing tasks.

These scenarios are based on our reading of the literature, personal experience, and personal interviews with people in the various fields we have studied who have displayed high levels of practical intelligence in their work (as determined through a nomination procedure). The interviewees agreed that their ability successfully to negotiate the work environment derived not from high IQ or

TABLE 11.1
Scenarios for Measuring Tacit Knowledge

Managing Self

You are concerned that you habitually put off completing disagreeable tasks and wish to improve this aspect of your work-related performance. Upon further examination, you think that your problem is one of procrastination--being unable to start tasks you need to get done on a given day. You have asked for advice about dealing with this problem from several friends in the company who seem to be especially productive when it comes to completing tasks. Rate the quality of the following pieces of advice that you have been given:

____ Wait to begin a given task until you really wish to do it.

____ Spend some time considering just what it is you dislike about a particular task and then try to change that aspect of it.

____ Get rid of all distractions (perhaps by taking the task into a conference room) so that there is nothing else you can do but the task you must complete.

____ Force yourself to begin the day by spending fifteen minutes on the task, in the hope that once you have started you will keep on it.

Managing Others

You have just learned that detailed weekly reports of sales-related activities will be required of employees in your department. You have not received a rationale for the reports. The new reporting procedure appears cumbersome and it will probably be resisted strongly by your group. Neither you nor your employees had input into the decision to require the report, nor in decisions about its format.

You are planning a meeting of your employees to introduce them to the new reporting procedures. Rate the quality of the following things you might do:

____ Emphasize that you had nothing to do with the new procedure.

____ Have a group discussion about the value of the new procedure and then put its adoption to a vote.

____ Give your employees the name and number of the director responsible for the new procedure, so that they may complain to that individual directly.

____ Promise to make their concerns known to your superiors, but only after they have made a good faith effort by trying the new procedure for six weeks.

____ Since the new procedure will probably get an unpleasant response anyway, use the meeting for something else and inform them about it in a memo.

____ Postpone the meeting until you find out the rationale for the new procedure.

Managing Tasks

You are responsible for selecting a contractor to renovate several large buildings. You have narrowed the choice to two contractors on the basis of their bids and after further investigation, you are considering awarding the contract to the Wilson & Sons Company. Rate the importance of the following pieces of information in making your decision to award the contract to Wilson & Sons:

____ The company has provided letters from satisfied former customers.

____ The Better Business Bureau reports no major complaints about the company.

____ Wilson & Sons' bid was $2000 less than the other contractors (approximate total cost of the renovation is $325.000).

____ Former customers whom you have contacted strongly recommend Wilson & Sons for the job.

Note: Examinees rate the quality of each piece of advice on a 1 (low) to 9 (high) scale.

even what they had learned in school, but from the work-related knowledge that they had picked up and exploited on the job—in other words, what we are calling *tacit knowledge*.

Having described now some of the basic elements of our construct of tacit knowledge and the theoretical framework into which it fits, we shall proceed to describe experiments with adults that have tested and/or expanded our conception of tacit knowledge.

EXPERIMENTS ON THE NATURE, USE, AND ACQUISITION OF TACIT KNOWLEDGE IN ADULTS

Experiment 1: Academic Psychologists

The following experiment was done as a collaboration between Richard Wagner and Robert Sternberg. The goal of this experiment was to construct-validate a theory and test of tacit knowledge for academic psychologists (Wagner & Sternberg, 1985). There were three groups of subjects.

Group 1 consisted of 54 members of the faculty in 20 psychology departments, either in the top fifteen by national rankings or not in the top fifteen. Group 2 consisted of 104 psychology graduate students sampled from the same departments as were the faculty. Group 3 consisted of 29 Yale undergraduates. Each subject received 12 work-related situations (4 for each type of tacit knowledge), each with from 6 to 20 response items. For example, one work-related situation described a second-year assistant professor who in the past year had published two unrelated empirical articles, who had one graduate student working with him, and who had as yet no external source of funding. His goal was to become a top person in his field and to get tenure in his department. Subjects had to rate on a 1 to 9 scale the value of several pieces of advice regarding what he could do in the next 2 months, given that he didn't have time to follow all of the pieces of advice. Examples of pieces of advice were to: (a) improve the quality of his teaching, (b) write a grant proposal, (c) begin a long-term research project that might lead to a major theoretical article, (d) concentrate on recruiting more students, (e) serve on a committee studying university-community relations, and (f) begin several related short-term research projects, each of which might lead to an empirical article.

The main independent variables in the study were tacit knowledge about managing oneself, others, and one's career, as well as group membership. The main dependent variable was the set of ratings to tacit-knowledge test items. As criterion variables against which to validate the tacit-knowledge test empirically, we obtained for the faculty members citation rates and numbers of publications from established reference sources, and we obtained from questionnaire data number of conferences attended within the last year, number of conference papers presented within the last year, distribution of time, academic rank, year of PhD, and level of institutional affiliation (higher or lower). For the undergraduates, we obtained scores on the Verbal Reasoning section of the Differential Aptitude Test (Bennett, Seashore, & Wesman, 1974).

Tacit-knowledge tests were scored by correlating ratings on each item with an index variable for group membership (3 = faculty member, 2 = graduate student, 1 = undergraduate). A positive item-group membership correlation would indicate that higher ratings were associated with more expertise, whereas a negative correlation would indicate the reverse. In a second analysis, we found significant positive correlations for the faculty members between tacit-knowledge scores and number of publications (.33), number of conferences attended (.34), rated level of institution (.40), and proportion of time spent in research (.39). We obtained significant negative correlations between tacit-knowledge test scores and proportion of time spent in teaching (−.29) and proportion of time spent in administrative activity (−.41). For the undergraduates who received the verbal reasoning test, there was no significant correlation between tacit-knowledge and verbal-reasoning scores (−.04). Other correlations were in the predicted direction. Thus, the tacit-knowledge test correlated well with at least some of the criteria against which it was validated, but did not correlate significantly with the standard test of verbal reasoning.

Experiment 2: Business Managers

The following experiment was done as a collaboration between Richard Wagner and Robert Sternberg. The goal of this experiment was to construct-validate the theory and test of tacit knowledge for business managers (Wagner & Sternberg, 1985). Again, we had three groups of subjects.

Group 1 consisted of 54 managers, 19 of whom were from among the top 20 companies in the Fortune 500, 28 of whom were not in these companies, and 7 who did not indicate their company affiliation. Group 2 consisted of 51 graduate students in five business schools varying in level of prestige. Group 3 consisted of 22 Yale undergraduates. Materials were twelve work-related situations, each with from 9 to 20 response items. The main independent variables were tacit knowledge about managing oneself, managing others, and managing one's career, as well as group membership. The main dependent variable was the set of ratings to the tacit-knowledge items. As criteria against which the tacit knowledge test could be validated for managers, we used level of company (top of the Fortune 500 or not in the Fortune 500), number of years of management experience, number of years of formal schooling, salary, number of employees supervised, and level of job title. Undergraduates took the Differential Aptitude Test (DAT), Verbal Reasoning subtest.

Again, we found some significant correlations for the professional group between the tacit-knowledge test and the criteria. Significant correlations were obtained for level of company (.34), number of years of schooling beyond high school (.41), and salary (.46). Other correlations were in the predicted direction. For the undergraduates, the correlation between tacit-knowledge scores and verbal-reasoning ability was not significant (.16), again suggesting that the tacit knowledge test was not merely a fancy conventional intelligence test.

Experiment 3: Bank Managers

The following experiment was done as a collaboration between Richard Wagner and Robert Sternberg. The goal of this third experiment was to cross-validate the management test on a new sample from a single company and occupation (Wagner & Sternberg, 1985).

Subjects were 29 managers from offices of a local bank. Materials were the same as in Experiment 2. The main independent variables were once again tacit knowledge regarding management of oneself, others, and one's career, whereas the dependent variable was the set of ratings to the tacit-knowledge items. Because we were now using managers from a single institution, it was possible to obtain more detailed criterion information. We obtained percentage of salary increase over the past 2 years, which in the bank was merit-based; overall performance ratings; ratings for managing personnel; ratings for generating new business; and ratings for implementation of bank policy and procedures.

We found significant positive correlations of the tacit-knowledge test with percentage of salary increase (.48), and with performance ratings for generating new business (.56), and implementation of bank policy and procedures (.39). Other correlations were in the predicted positive direction. Thus, the test successfully cross-validated to a new sample.

Experiment 4: Academic Psychologists II

The following experiment was part of a dissertation done by Richard Wagner under the supervision of Robert Sternberg. The goal of this experiment was to construct-validate a revised version of the theory of tacit knowledge as well as a revised version of the test (Wagner, 1987). The new theory and test separated global versus local tacit knowledge, and also distinguished between people's conceptions of real versus ideal jobs. In other words, people could have conceptions of what to do in the job they actually had, or they could have conceptions of what to do in the ideal job. The question was whether both of these conceptions would correlate with job performance.

Subjects were again divided into three groups. Group 1 consisted of 91 faculty members in 26 departments of psychology that ranged in rated scholarly quality. Group 2 consisted of 61 graduate students from the same departments. Group 3 consisted of 60 Yale undergraduates. All subjects received twelve work-related situations with from 9 to 11 response items. Ratings for both actual and ideal jobs were to be given. The main independent variables in the experiment were three types of ratings (management of oneself, others, and tasks) crossed with two orientations of such ratings (local tacit knowledge and global tacit knowledge). Dependent variables were ratings for actual and for ideal jobs. Criterion variables were rated quality of department, number of citations, and number of publications for faculty members, and DAT Verbal-Reasoning scores for undergraduates.

In this experiment, a new scoring method was used. A sample of highly practically-intelligent professors was obtained through a nomination process, and tacit-knowledge tests were scored in terms of the distance (d^2) of each individual profile from the expert profile. Whereas in the previous experiment mean scores on the tests would not have been meaningful because the scoring procedure was designed to discriminate between groups, in this experiment mean differences were meaningful.

Mean d^2 values for the three groups were 339 for faculty, 412 for graduate students, and 429 for undergraduates, indicating that, on the average, tacit knowledge increased with level of experience. Of course, there were exceptions within groups, indicating that what mattered was not merely experience but what one has learned from experience.

Significant correlations were obtained with the criterion variables. Note that now a negative correlation indicates an association between better tacit-knowledge scores and the criteria, because with the distance measure, a better tacit-knowledge score is a lower distance score. For the actual-job ratings, significant correlations were obtained for the faculty members with ratings of department (−.48), number of citations (−.44), number of publications (−.28), proportion of time spent on research (−.41), and number of papers presented (−.21). Significant positive correlations, indicating an association between higher criterion variables and lesser tacit knowledge, were obtained between the tacit-knowledge scores and proportion of time spent in teaching (.26) and proportion of time spent in administration (.19). For the ideal ratings, correlations were generally slightly lower but comparable. Significant correlations were obtained for rating of department (−.42), number of citations (−.43), and proportion of time spent in research (−.34). We also looked at the intercorrelations of the various scales. The six scales were ratings of (a) oneself-local, (b) oneself-global, (c) others-local, (d) others-global, (e) task-local, and (f) task-global. We found that 13 of the 15 intercorrelations were significant and positive, with correlations generally in the .2 to .4 range. The two nonsignificant correlations were both in the positive direction. These correlations indicated at least weak "g" (general cognitive factor) for tacit knowledge. In other words, people who scored higher on one of these subscales tended to score higher on the others as well. Thus, although tacit-knowledge scales do not correlate with verbal-reasoning ability, the subscales of the tacit knowledge scales do correlate with each other: People who are higher in one aspect of tacit knowledge also tend to be higher in others.

Experiment 5: Business Managers II

The following experiment was part of a dissertation done by Richard Wagner under the supervision of Robert Sternberg. The goal of this experiment was to construct-validate the revised tacit-knowledge theory and test for global and local tacit knowledgge and for real and ideal jobs with business managers (Wagner, 1987).

Again, there were three groups of subjects. Group 1 consisted of 64 business managers from 31 companies. Of these managers, 26 were from companies in the top forty of the Fortune 500 list; 33 were from companies not on the Fortune 500 list; and 5 were from companies whose identity was not indicated. Group 2 consisted of 25 business graduate students from 7 business schools of varying quality. Group 3 consisted of 60 Yale undergraduates. The main materials were twelve work-related situations. The independent variables were the three contents (ratings of self, others, and tasks) crossed with the two orientations of ratings (local, global). Dependent variables were ratings for actual and for ideal jobs. Criterion variables were salary, number of years of management experience, level of company, and number of years of formal schooling beyond high school.

Scoring was again done via a distance measure from the prototype of an expert group. Mean scores were 244 for the business managers, 340 for the business graduate students, and 417 for the undergraduates, again indicating greater tacit knowledge as a function of experience. Correlations with criterion variables were somewhat lower than in the previous experiments. For the actual ratings, a significant correlation was obtained for number of years of management experience (−.30). Other correlations were in the predicted direction but not significant. For the ideal ratings, significant correlations were obtained with salary (−.32) and with number of years of management experience (−.27). The correlation with level of company was in the predicted direction, that with years of schooling beyond high school was not (although it was not significant in the other direction). Once again, these six subscales were generally significantly correlated with each other. Twelve of fifteen possible correlations were significant, with values ranging from the .2 level to the .5 level. The other three correlations were in the expected direction. Thus, once again, there was an appearance of a general factor, albeit a weak one, for tacit knowledge.

In this experiment, the same undergraduates took both the academic-psychology and the business-management tacit-knowledge tests. The correlation between scores on the two tests was .58, which was highly significant; thus, it appeared that not only do the subscales of the tacit-knowledge test correlate with each other, but so do scores on two different tests of tacit knowledge. In other words, there appears to be at least some common core of tacit knowledge between disciplines, although the correlation is not high enough to indicate that the tacit knowledge is the same. Rather, there appears to be some tacit knowledge that is general, and some that is not.

Experiment 6: Center for Creative Leadership LDP Business Managers

This experiment was a collaboration between Richard Wagner and Robert Sternberg. The goals of this experiment were to construct-validate the tacit-knowledge test against behavior in a managerial simulation, and also to study the incremental validity of the test. In the previous experiments, our criteria were all static. In this experiment, we were able to obtain as a criterion performance in a managerial simulation, which is a more dynamic kind of assessment. We were also able to obtain scores on a wide variety of psychological measures, so that it was possible to determine whether tacit knowledge qualitatively differed from kinds of attributes measured in psychological tests beyond simply the verbal-reasoning ability that we had measured in the prior experiment.

Subjects were 45 participants in the Leadership Development Program (LDP) at the Center for Creative Leadership in Greensboro, North Carolina. Participants were generally mid- to upper-level executives. Materials were nine work-related scenarios, each with ten response items. In addition, we had available from the Center for Creative Leadership scores on a number of psychological tests. Tests included a test of intelligence (the Shipley), the California Psychological Inventory, the Myers-Briggs Type Indicator, the Fundamental Interpersonal Relations Orientation-Behavior (FIRO-B), the Hidden Figures Test, the Kirton Adaptation-Innovation Inventory, a managerial job satisfaction questionnaire, and behavioral assessment data from two managerial simulations. The independent variables were the various intellectual and personality tests. The dependent variable was performance on the managerial simulations.

The main question was whether in predicting performance on the managerial simulation, the tacit-knowledge test showed insignificant statistically significant incremental prediction (ΔR^2) over and above the prediction of other measures. Quite simply, the results were uniformly favorable for the tacit-knowledge test. Values of ΔR^2 for the tacit-knowledge test in predicting the simulation scores were .32 beyond IQ, .22 beyond CPI combined with IQ, .32 beyond the FIRO-B combined with IQ, .28 beyond field independence combined with IQ, .33 beyond innovation scores combined with IQ, .35 beyond the Myers-Briggs combined with IQ, .32 beyond job satisfaction combined with IQ, and .17 beyond all five predictors reliably individually correlated with the simulation. In other words, even with all reliable predictors entered, the tacit knowledge test still contributed incrementally. All foregoing values of ΔR^2 were statistically significant. Thus, the tacit-knowledge test appears to measure a new construct, not just to rehash other constructs already in the psychological literature.

Experiment 7: Salespeople

This experiment was done as a collaboration among Richard Wagner, Carol Rashotte, and Robert Sternberg. Our goal in this experiment was to construct-validate a "rules-of-thumb" approach to the understanding and measurement of tacit knowledge in salespeople. In all of the previous experiments, scoring was empirically derived, whether from the directions of correlations between items and an index for group membership, or from d^2 comparison between expert and experimental-protocols. We believed that it would be preferable to have a more objective, expert-based scoring scheme for evaluating performance on the tacit-knowledge tests. In our work in sales, therefore, we decided to seek the rules of thumb that sales people use in order to optimize their performance. Through interviews, reading of books on sales, and their own experience, we generated a list of rules of thumb. The rules of thumb were divided into several main categories. For example, one such category was setting sales goals. Examples of rules of thumb under this category were: (a) target sales goals in number of units sold, not dollars; (b) set goals that are measurable and specific; and (c) commit to reaching your sales goals in writing. Another category was handling the customer who stalls. Examples of rules of thumb here would be: (a) play your hunches and ask if you suspect a competitor has entered the picture; and (b) penetrate smoke screens by asking "what if . . .?" questions. Another category, attracting new accounts, would have as examples of rules of thumb (a) be selective in regard to whom you direct your promotion efforts, and (b) ask your customers to provide leads. Or handling the competition, another category, would include as rules of thumb (a) build up your product and company rather than tear down your competitor's; and (b) remember that customers buy for their reasons, not yours.

Subjects in the experiment were divided into two groups. Group 1 consisted of 30 salespeople (who sold automobiles, furniture, or houses) with an average of 14 years of selling experience. Group 2 comprised 50 undergraduates at Florida State University. Each subject received eight work-related scenarios, with 8 to 12 response items constructed via a rules-of-thumb approach. The construction of the items was such that some of them accurately represented the rules of thumb, whereas others represented weakened or slightly distorted versions of them. We could therefore see the extent to which subjects preferred the items that represented the actual rules of thumb versus the extent to which they preferred items that represented distorted versions of these rules. The Differential Aptitude Test, verbal reasoning subscale was administered to the undergraduates. An example of a tacit-knowledge item for sales appears in Table 11.2.

The main independent variables in the experiment were local versus global tacit knowledge, and membership in the salesperson versus undergraduate

TABLE 11.2
Example of Tacit Knowledge Item for Sales

You sell a line of photocopy machines. One of your machines has relatively few features and is inexpensive, at $700, although it is not the least expensive model you carry. The $700 photocopy machine is not selling well and it is overstocked. There is a shortage of the more elaborate photocopy machines in your line, so you have been asked to do what you can to improve sales of the $700 model.

Rate the following strategies for maximizing your sales of the slow-moving photocopy machine.

A. Stress with potential customers that although this model lacks some desirable features, the low price more than makes up for it.

B. Stress that there are relatively few models left at this price.

C. Arrange as many demonstrations as possible of the machine.

.....

J. Stress simplicity of use, since the machine lacks confusing controls that other machines may have.

Note. Subjects rated items on a 1 (low) to 9 (high) scale.

groups. The main dependent variable was the set of responses to the tacit-knowledge test.

We found that scores on the tacit-knowledge test improved with experience for both local and global tacit knowledge. For local tacit knowledge, the mean for salespeople was 99, versus 74 for undergraduates. For global tacit knowledge, the mean for salespeople was 110 versus 92 for the undergraduates. The total for salespeople was therefore 209 versus 166 for the undergraduates. Thus, people scored higher on the measure with more experience in sales. We also found that whereas global tacit knowledge did not correlate (.05) with the Differential Aptitude Test verbal reasoning subsection, local tacit knowledge did (.40). For the first time, then, we obtained a significant correlation between tacit-knowledge scores and verbal reasoning ability, but only for local tacit knowledge.

Experiment 8: Salespeople II

This experiment was done as a collaboration among Richard Wagner, Carol Rashotte, and Robert Sternberg. The goal of the experiment was the external validation of the tacit-knowledge theory and test for salespeople with measures of actual performance in sales.

Subjects were divided into two groups. Subjects in the first group comprised 48 life-insurance salespeople with an average of 11 years of selling experience. Subjects in the second group consisted of 50 undergraduates at Florida State University with no sales experience. The main materials were the tacit-knowledge measure for sales (from Experiment 7) and, for undergraduates, the

DAT verbal-reasoning test. The main independent variables were local versus global tacit knowledge, and membership in the salespeople versus undergraduate groups. The main dependent variable was the set of tacit-knowledge scores. We also had criterion variables against which to validate the tacit knowledge test: number of years with the company, number of years in sales, number of yearly quality awards, 1985 sales volume, 1986 sales volume, 1985 premiums, 1986 premiums, college background, and business education.

Once again, tacit knowledge increased with level of experience. The respective scores for undergraduates and salespersons were, for local tacit knowledge, 73 and 94; for global tacit knowledge, 92 and 112; and for total score, 165 and 206. Thus, again, salespersons did better than did undergraduates. We found some correlations between our tacit-knowledge test and the criterion variables. For total score, significant correlations were obtained with number of years with the company (.37), number of years in sales (.31), number of yearly quality awards (.35), and business education (.41). However, it turns out that local and global tacit knowledge did not contribute equally to these correlations. For local tacit knowledge, significant correlations were obtained with only three of the variables: number of years with the company (.23), number of yearly quality awards (.28), and business education (.35). For global tacit knowledge, however, significant correlations were obtained with seven criterion variables: number of years with company (.32), number of years in sales (.28), number of yearly quality awards (.25), 1985 sales volume (.37), 1986 sales volume (.28), 1985 premiums (.26), and 1986 premiums (.29). For the undergraduates, once again, local tacit knowledge correlated significantly with the DAT verbal (.25), whereas global tacit knowledge did not (−.02). Thus, the kind of tacit knowledge that correlated significantly with the ability test showed substantially lower correlations with the sales criteria than did the kind of tacit knowledge that did not correlate with the ability test. Once again, then, practical intelligence seems to be something qualitatively different from the academic intelligence as measured by conventional tests.

Experiment 9: Tacit Knowledge in College Students

The following experiment was done as a collaboration between Wendy Williams and Robert Sternberg. In a prestudy, 50 Yale college students replied to the question: "What does it take to succeed at Yale that you don't learn from textbooks?" The results of this prestudy were used to form a tacit-knowledge inventory for college students.

Our main goals in the experiment were (a) to identify the tacit knowledge necessary for success as a college student, (b) to compare tacit knowledge for students at different points in their college careers, (c) to assess the role of tacit knowledge in college-student success, (d) to begin construct validation of a tacit-knowledge inventory for college students.

Subjects in this experiment were 53 Yale college students. Of these subjects, 18 were male and 35 were female. They were divided among classes: 18 freshmen, 3 sophomores, 9 juniors, and 23 seniors.

The main materials were the tacit knowledge test for college students. It consisted of fourteen situations, each with associated response options to be rated on a 1–9 scale.

The items were similar in spirit, but different in content from those described earlier. For example, one described the subject as enrolled in a large introductory lecture course. The requirements consisted of three term-time exams and a final. Subjects were asked to rate how characteristic it was of their behavior to spend time doing various activities, such as (a) attending class regularly, (b) attending optional weekly review sections with a teaching fellow, (c) reading assigned text chapters thoroughly, (d) taking comprehensive class notes, and (e) speaking with the professor after class and during office hours. Another example of an item would require students to rate how important they believed the average professor considered various activities to be for a student. Examples of such activities would be (a) making an effort to speak with the professor before or after class, (b) meeting with the professor during office hours, (c) completing work ahead of schedule—handing work in early, (d) attending class regularly and arriving on time, (e) writing especially creative and unusual papers, (f) getting high grades on exams, and (g) getting high grades on papers.

The criterion measures in this study were two main indices: an academic index and an adjustment index. The academic index was a composite of high school GPA, college CPA, SAT scores, and CEEB achievement test scores. The adjustment index was a composite of a measure of happiness in college, a measure of self-perceived success in college, a measure of self-perceived success in using tacit knowledge, a measure of the extent of benefit each subject had experienced from learning tacit knowledge, and a measure of the rated closeness of the college to the subject's ideal college. The main independent variables were tacit knowledge scores, year in college, and gender. The main dependent variables were the composite academic and adjustment indices.

The academic and adjustment indices, the main dependent variables in the study, were uncorrelated (−.09). Perhaps embarrassingly, the correlation between year at Yale and the adjustment index was negative and significant (−.43). Thus, subjects' self-perceived adjustment declined with number of years in the college.

A number of items showed significant correlations with the academic index: perceived importance of maintaining a high GPA (.42), doing extra reading and school work not specifically assigned (.27), not attending optional weekly review sections with a teaching fellow (.32), not skimming required reading the morning before class (.37), not preparing a brief outline of points to raise in class discussion (.31), not helping friends with their assignments (.34), not behaving consistently from situation to situation (.25), finding it uncharacteristic to accept pressure and stress as parts of life (.30), finding it uncharacteristic to stand up for oneself (.34), and its being uncharacteristic to play a sport or exercise regularly (.45).

In general, a different set of items correlated significantly with the adjustment index: beliefs that professors value a clear, direct writing style, good organization of thoughts and ideas, and creative or unusual ideas (.38); beliefs that professors value papers that bring in outside interests or material (.27); beliefs that it is important sometimes to take on too many responsibilities at once (.31), seeking advice from several faculty in addition to one's own professors (.31), taking classes that permit occasional absences (.36), being positive and looking on the bright side of life (.42), not being intimidated (.33), being flexible (.27), maintaining a strong sense of confidence and independence (.37), not worrying unnecessarily or destructively (.31), knowing how to make oneself happy (.32), and not letting small disappointments affect one's long-term goals (.29).

We obtained some rather interesting gender differences. Males rated higher than females on three items: belief that professors value funny and entertaining papers, downplaying the seriousness of cheating, and worrying less. Females rated higher than males: taking comprehensive class notes, believing that professors value papers that express special interests and enthusiasms, trying to figure out what makes them happy, thinking about what they are able to do best, trying to discover and understand limitations, and cultivating a sense of responsibility and commitment. The results were very much in line with Gilligan's (1982).

Comparing items that freshmen rated higher than seniors with items that seniors rated higher than freshmen, one might conclude that students become somewhat more cynical over time. Freshmen rated higher than seniors items such as believing that professors value papers with no typographical or grammatical errors, with creative ideas and unusual ideas, with mention of outside interests, with accurate and thorough references, and with a demonstration of effort and motivation. Seniors rated higher than freshmen items that reflect the beliefs that professors value students that get high grades and that they are likely to ask forgiveness if they are caught cheating.

Using relatively small sets of items from the tacit-knowledge scale, we were able to obtain fairly good prediction of both the academic and the adjustment indices. With four items, the overall R^2 with the academic index was .43. The four items were: not preparing an outline of points to raise in class discussion, maintaining a high GPA, not helping friends with assignments, and not playing a varsity or intramural sport. For the adjustment index, the R^2 was .63. The six items contributing significantly to this prediction were: believing professors value a clear, direct writing style; maintaining a strong sense of confidence and independence; standing up for oneself; sometimes taking on too many responsibilities at once; seeking advice from faculty in addition to the instructor of the course; and taking classes that permit occasional absences. In sum, then, tacit knowledge predicts both academic performance and adjustment in college, and thus is important not only in occupational settings, but in school settings as well. To succeed in school, one needs not only formal knowledge, but informal or tacit knowledge.

Experiment 10: Acquisition of Tacit Knowledge

This experiment was done as a collaboration among Lynn Okagaki, Robert Sternberg, and Richard Wagner. The goal of the experiment was to demonstrate that tacit knowledge is acquired through three knowledge-acquisition components specified by the componential subtheory of Sternberg's (1985) triarchic theory of human intelligence: selective encoding, selective combination, and selective comparison. Selective encoding involves distinguishing relevant from irrelevant information in the course of learning new material. Selective combination involves putting the relevant information together in order to form a whole, unified cognitive structure. Selective comparison involves drawing upon past information relevant to the present in order to facilitate learning of new information.

Subjects were divided into five groups: two control groups and three experimental groups. In Group C_1, a control group, 15 college students received as a pretest and a posttest the tacit-knowledge measure for sales, with no intervening treatment. In Group C_2, a second control group, 15 college students received the tests as well as a tacit-knowledge acquisition task with no cuing to help them identify or use relevant information. In Group E_1, an experimental group, 15 college students received the tests and also the acquisition task with selective-encoding cuing. Relevant information for acquisition of tacit knowledge was highlighted and the relevant rule-of-thumb provided. In Group E_2, another experimental group, 15 college students received the tests and also the acquisition task with selective-combination cuing. Relevant information was highlighted, the relevant rule-of-thumb was given, and a note-taking sheet with the appropriate categories was given to subjects in order to help them combine information. In Group E_3, the third experiment group, 15 college students received the tests and also the acquisition task with selective comparison cuing. This cuing was an evaluation that had been completed by a "predecessor" in the company. The idea was that they could use the predecessor's performance to facilitate their own. Relevant information was highlighted and rules-of-thumb were also given to people in this group.

The main materials in the experiment were the tacit-knowledge test for sales and the tacit-knowledge acquisition task. In the task, subjects took the role of a personnel manager whose immediate job was to read the transcripts of three job interviews with applicants for sales positions in his company. The experimenter asked the subjects to evaluate the applicants' ability to manage themselves, to handle the tasks and problems that arise in sales positions, and to handle business relationships with customers, peers, and superiors (i.e., managing oneself, managing tasks, and managing others). The subjects received evaluation forms on which they rated each applicant on each of the categories, gave an overall rating of the applicant, and indicated whether or not the applicant should be hired. In addition, subjects were asked to identify all sentences they had used in the interview protocol in making their evaluations, to indicate the category of information that was relevant (managing oneself, managing tasks, managing others), and to indicate whether the information in each sentence was positive or negative with respect to their decision.

The main independent variable in the experiment was group assignment. The main dependent variables were scores on the acquisition task and difference scores for the tacit-knowledge posttest minus the tacit-knowledge pretest. On the acquisition task, there were three types of scores: (a) hit, that is, the total number of relevant sentences identified as relevant (plus an additional point for each correct assignment of positivity versus negativity of the information); (b) miss, that is, the total number of relevant sentences not correctly identified as such; and (c) false alarm, that is, the total number of relevant sentences identified as relevant. Subjects were instructed to take the role of the manager who is evaluating three possible candidates for sales positions in the company. The manager can hire none, one, two, or all of the candidates. The important thing is to hire only those persons who have the most potential for being good salespeople in the company. The subjects were first given a two-page description of the company to read. Then they were given transcripts of the three job interviews. They were instructed to read through all three interviews before they did their evaluations.

In the tacit-knowledge acquisition task, we did a Manova with the experimental group as the independent variable and hits, misses, and false alarms as the dependent variables and found that there was an overall difference among the four groups (C_2, E_1, E_2, and E_3). The second control group (C_2) performed significantly worse than the three experimental groups. The mean number of hits was lowest in C_2 (27.7), and the number of misses and false alarms was highest in this group (90.5 and 54.3 respectively). Among the experimental groups, the selective-combination group did the best: 70.2 hits, 49.9 misses, 21.3 false alarms. Performance of the selective-encoding and selective-comparison groups was comparable, and lower than the selective-combination group but higher than the control group: 45.5 hits, 74.7 misses, and 28.3 false alarms for the selective-encoding group (E_1), and 41.1 hits, 78.8 misses, and 19.8 false alarms for the selective-comparison group (E_3).

We also looked at posttest minus pretest difference scores on the tacit-knowledge tests for all five groups. Group C_1 (the control group with no experimental task) showed the least gain, with a mean difference score of 3.5. Group C_2 did better, with a mean of 7.7. Group E_3, the selective-comparison group, did slightly better at 9.3, followed by Group E_1 at 16.8 and E_2 at 19.7. Thus, C_1 was the worst, C_2 marginally worse than the other groups, and E_1 and E_2 better than the rest. These results suggest that the selective-comparison manipulation was the weakest of the three, but that the selective-encoding and selective-combination manipulations were quite successful in inducing learning of tacit knowledge for sales.

TEACHING PRACTICAL INTELLIGENCE FOR SCHOOLS

The experiments described so far have all been done with adults, where the goal has been to understand the nature, use, and acquisition of tacit knowledge. Children also need tacit knowledge, especially in a school setting, where there are many unspoken rules as to what kinds of behavior are acceptable and what kinds are not, and as to how one is supposed to work and interact in the school environment.

In the experiment described here, we seek to demonstrate that practical intelligence can be taught. Our subjects are children rather than adults, and the practical intelligence that they are taught is that relevant to a school setting. The experiment described below represents a collaboration among Robert Sternberg, Lynn Okagaki, and Alice Jackson (Sternberg, Okagaki, & Jackson, 1990). The project as a whole is being done as a collaboration between Yale University and Harvard University, with the Harvard team led by Howard Gardner. We describe here only the Yale portion of the project, however. The goal of the experiment is described first, then its theoretical basis, the instructional model underlying it, the form and content of it, the data from a study done in a middle-class suburb in Connecticut, and finally we draw some conclusions.

Introduction to the Practical-Intelligence-For-School (PIFS) Program

The goal of the Practical Intelligence For School (PIFS) program is to teach children how to apply their intelligence effectively in a school setting. The emphasis is on both the acquisition and the application of school-survival skills. Teachers generally do not explicitly teach such skills, such as how to study for different kinds of tests, how to allocate time in doing homework, how to decide what kinds of things can and cannot be said to teachers, figuring out how to get along with classmates, and so on. Teachers generally assume that this is not material that they should teach, or that the students have learned the material in the past. However, many students never learn these practical skills, with the result that they are at a competitive disadvantage in the school situation. They may, in fact, acquire the formal knowledge for success in school, but without the informal or tacit knowledge, perform at a level that does not reflect their true ability.

The theoretical basis for the PIFS program is the process orientation of the triarchic theory of human intelligence (Sternberg, 1985) combined with the theory of multiple intelligences (Gardner, 1983). The processes of the triarchic theory are infused into the content domains of the theory of multiple intelligences so as to instigate learning of how to apply practical intelligence for school in each of the seven domains of linguistic, logical-mathematical, spatial, musical, bodily-kinesthetic, interpersonal, and intrapersonal intelligences. Thus, students learn how to manage themselves, manage tasks, and work with others in each of seven domains.

The PIFS program consists of two main parts: a core course relevant to all subject matter areas and infusion units relevant to individual subject-matter areas. The Yale group has been responsible for the core course, which will be described here. This course itself consists of a student guide, a teacher guide, and a set of classroom experiences.

Four mechanisms for achieving transfer were built into the program (Sternberg & Frensch, 1992). The first mechanism is encoding specificity (Tulving & Thomson, 1973). Material is taught in a way such that it will be used in the same manner that it is encoded. In other words, the material is taught in the context of practical situations that are similar or identical to the kinds of practical situations students will encounter in school. Second, the material is organized in a way so that it can be retrieved when it is later needed. Third, students learn how to discriminate the various kinds of material—when to use which of the specific techniques they learn. Finally, students are encouraged to develop a set for using the material they learn so that when they are in actual school settings, they will use the information, rather than its remaining encapsulated in the context of the course in which it was taught.

In the study described here, the course lasted for one semester. However, only about one-half of the material was completed, whereas ideally, the course should be taught for 1 year. It met three times per week for roughly 45 to 50 minutes per session, and was done in three separate reading classrooms at the 7th-grade level.

The program was divided into three main units: managing yourself, managing tasks, and cooperating with others. Under the managing yourself unit, the main topics were: kinds of intelligence, understanding test scores, what's your style?, taking in new information, showing what you learned, knowing how you work best, recognizing the whole and the parts, memory, using what you already know, making pictures in your mind, using your eyes—a good way to learn, try doing it!, accepting responsibility, and collecting your thoughts—setting goals. Under managing tasks, the main topics were getting organized, is there a problem?, what strategies are you using?, a process to help you solve problems, planning a way to prevent problems, breaking habits, help with our problems, thinking about time, understanding questions, following directions, and taking tests. Under cooperating with others, the main topics were: class discussions, what to say during discussions, tuning your conversation, putting yourself in another's place, solving problems in communication, making choices—

adapting, shaping, and selecting, and seeing the relationship between now and later.

Suppose, for example, that a problem in getting along with others was a student's realization that he is not popular enough. One set of solutions for solving the problem would be to solve it as stated. The student might become more willing to do what others are interested in doing, or learn not to one-up others, or learn to listen to others, or learn to see things from others' points of view. Or the student might learn how to redefine the problem. For example, he might redefine what is meant by "enough." Perhaps he should set his sights lower or should reduce the importance of being popular—just how important is it to him to be popular? Or he might ask himself: "With whom?" He might find a new social group with whom he might be more popular, or he might redefine what he means by "popular." He could also ask himself, "Popular with respect to what?"

EXPERIMENT 11: AN IMPLEMENTATION OF THE PIFS PROGRAM

Subjects were 110 7th-grade students at a middle school in central Connecticut. The students were almost balanced for gender. Of the total, 61 were experimental subjects who received the PIFS program whereas 39 were controls who did not. Three seventh-grade teachers taught the course.

Materials were of two main kinds. The first kind was the course itself, comprising the students' guide, the teachers' guide, and classroom experiences. In addition, we used three tests for pretest-posttest evaluation. The first test was the Survey of Study Habits and Attitudes (Brown & Holtzman, 1967). The second test was the Learning and Study Skills Inventory (Weinstein & Palmer, 1988). Both of these tests measured skills of the kinds that should have been taught directly in the course, although the course was designed without consultation of these tests. The third test, used to measure transfer, was the practical-contextual section of the Sternberg Triarchic Abilities Test (Sternberg, in preparation). Experimental subjects received the pretest, followed by instruction, followed by the posttest. The controls received only the pretest and the posttest.

The three classes met during reading period 3 days per week for about 45 to 50 minutes during the spring, 1989 semester. The controls also took the test during reading classes, and instead of receiving the PIFS program received reading instruction during the time of the program. Teachers were extensively pre-serviced and in-serviced. In other words, we spent about 3 days with them before the course in order to teach them how to implement it, and then continued to meet with them during the time that the course was actively meeting. Implementation of the program was not straightforward. As we went along, we continued to discuss better ways of doing things and new ways of conveying the content of the lesson. Moreover, in some cases, it was necessary to help the teachers adopt an attitude whereby they would be teaching primarily for thinking rather than primarily for merely conveying course contents.

A multivariate analysis of variance was done in order to determine whether there was an interaction between group membership and posttest minus pretest difference scores for each of the subscales within each of the three measures (see Okagaki & Sternberg, in press). As the interactions were statistically significant for each of the three tests, we proceeded with the analysis.

For the Survey of Study Habits and Attitudes, the experimental group gained significantly more than the control group for each of the four subscales: delay avoidance, work methods, teacher approval, and education acceptance. For the Learning and Study Skills Inventory, all of the posttest minus pretest differences were in the predicted direction, and eight out of ten were significant. The significant differences in favor of the experimental group were for the subscales of attitude, motivation, anxiety, concentration, information processing, selecting main ideas, self-testing, and test strategies. Nonsignificant differences (in the predicted direction) were for time management and study aids. For the Sternberg Triarchic Abilities Test, statistically significant differences favoring the experimental group were obtained for the practical inference (verbal) and route-planning (figural) subtests. The one result among all of those considered that went in the direction opposite to that predicted was for the practical-data (quantitative) subtest, where the controls gained significantly more. We are unable to account for this one unpredicted result.

CONCLUSIONS

To summarize our findings, several main points emerged from our studies.

First, the concept of tacit knowledge is important for understanding work performance in multiple domains, such as academic psychology, business management, sales, and even academic study. In some cases, global versus local tacit knowledge predict reasonably well. Second, tacit knowledge increases, on the average, with amount of experience in a domain, but it is what is learned from experience, not the experience itself, that matters. There can be wide individual differences within groups of comparable levels of experience. Third, tacit-knowledge scores correlate poorly, if at all, with conventional ability-test scores, at least within the ranges of subjects we tested. It is important to remember, though, that the people who actually go into the occupations we studied do not represent the full range of possible intelligence-test scores, but rather a truncated and above-average range. Fourth, tacit knowledge is also not a proxy for conventional measures of personality. Fifth, tacit-knowledge scores correlate moderately with each other and with external criteria performance. One apparently can predict tacit knowledge in one area from tacit knowledge in another, and one can obtain moderate prediction to external criteria of success. Of course, these criteria are ones that are defined by the field as a whole, and do not necessarily correspond to the criteria of success of a given individual. Sixth, the knowledge-acquisition components of selective encoding and selective combination, and probably selective comparison, are important to the acquisition of tacit knowledge. Seventh, tacit knowledge, and its associated rules of thumb, are by no means all that matter for job performance. We need to go beyond conventional tests to understand what else does matter. Eighth, tacit knowledge can be taught.

In sum, our studies show that tacit knowledge is important to success, and is not merely a fancy proxy for IQ. The goal of ability testing has always been to assess a person's ability to adjust in the world, not only the world of school, but the world of work. Testing tacit knowledge provides a unique entree into assessing adjustments of both kinds.

ACKNOWLEDGMENTS

The research with adults was supported by a contract from the Army Research Institute. The research with children was supported by a grant from the McDonnell Foundation. We are grateful to Wendy Williams, Carol Rashotte, and Alice Jackson for their collaborations in specific experiments described in this chapter.

REFERENCES

Bennett, G. K., Seashore, H. G., & Wesman, A. G. (1974). *Differential Aptitude Tests* (Form T). New York: The Psychological Corporation.

Brown, W. F., & Holtzman, W. H. (1967). *Survey of study habits and attitudes* (Form H). (SSHA) New York: The Psychological Corporation.

Ceci, S. J., & Liker, J. (1986). Academic and nonacademic intelligence: An experimental separation. In R. J. Sternberg & R. K. Wagner (Eds.), *Practical intelligence: Nature and origins of competence in the everyday world* (pp. 119–142). New York: Cambridge University Press.

Chi, M. T. H., Glaser, R., & Farr, M. (Eds.). (1988). *The nature of expertise.* Hillsdale, NJ: Lawrence Erlbaum Associates.

Gardner, H. (1983). *Frames of mind: The theory of multiple intelligences.* New York: Basic Books.

Gilligan, C. (1982). *In a different voice.* Cambridge, MA: Harvard University Press.

Lave, J., Murtaugh, M., & de la Rocha, O. (1984). The dialectic of arithmetic in grocery shopping. In B. Rogoff & J. Lave (Eds.), *Everyday cognition* (pp. 67–94). Cambridge, MA: Harvard University Press.

Okagaki, L., & Sternberg, R. J. (in press). Putting the distance into students' hands: Practical intelligence for school. In R. R. Cocking & K. A. Renninger (Eds.), *The development and meaning of psychological distance.* Hillsdale, NJ: Lawrence Erlbaum Associates.

Polanyi, M. (1976). Tacit knowledge. In M. Marx & F. Goodson (Eds.), *Theories in contemporary psychology* (pp. 330–344). New York: Macmillan.

Scribner, S. (1986). Thinking in action: Some characteristics of practical thought. In R. J. Sternberg & R. K. Wagner (Eds.), *Practical intelligence: Nature and origins of competence in the everyday world* (pp. 13–30). New York: Cambridge University Press.

Sternberg, R. J. (1985). *Beyond IQ: A triarchic theory of human intelligence.* New York: Cambridge University Press.

Sternberg, R. J. (in preparation). Sternberg Triarchic Abilities Test. San Antonio, TX: The Psychological Corporation.

Sternberg, R. J., & Frensch, P. A. (1992). Mechanisms of transfer. In D. K. Detterman & R. J. Sternberg (Eds.), *Transfer on trial: Intelligence, cognition, and instruction.* Norwood, NJ: Ablex.

Sternberg, R. J., Okagaki, L., & Jackson, A. (1990). Practical intelligence for school: A plan and a program for teaching school survival skills. *Educational Leadership, 48,* 35–39.

Sternberg, R. J., & Wagner, R. K. (Eds.). (1986). *Practical intelligence: Nature and origins of competence in the everyday world.* New York: Cambridge University Press.

Tulving, E., & Thomson, D. M. (1973). Encoding specificity and retrieval processes in episodic memory. *Psychological Review, 80,* 352–373.

Wagner, R. K. (1987). Tacit knowledge in everyday intelligent behavior. *Journal of Personality and Social Psychology, 52,* 1236–1247.

Wagner, R. K., & Sternberg, R. J. (1985). Practical intelligence in real-world pursuits: The role of tacit knowledge. *Journal of Personality and Social Psychology, 48,* 436–458.

Wagner, R. K., & Sternberg, R. J. (1986). Tacit knowledge and intelligence in the everyday world. In R. J. Sternberg & R. K. Wagner (Eds.), *Practical intelligence: Nature and origins of competence in the everyday world* (pp. 51–83). New York: Cambridge University Press.

Weinstein, C., & Palmer, D. (1988). *Learning and Study Skills Inventory.* Clearwater, FL: H & H Publishing Co.

Wigdor, A. K., & Garner, W. R. (Eds.). (1982). *Ability testing: Uses, consequences, and controversies.* Washington, DC: National Academy Press.

V SUMMARY AND INTEGRATION

12 A Survey of Research in Everyday Cognition: Ten Views

James M. Puckett
Texas A & M University-Kingsville

Leslee K. Pollina
Joseph S. Laipple
Ruth H. Tunick
Frank H. Jurden
West Virginia University

The purpose of this chapter is to provide an integrated summary of each of the ten primary chapters in this volume and, in doing so, identify common themes, differences in empirical findings and opinions, and theoretical connections between them. Puckett, Reese, and Pollina (this volume) address four themes that are used to help organize this integration, so we summarize that chapter as well.

First Puckett et al. provide a brief historical overview of the continuing debate over controlled vs. naturalistic methods (e.g., Banaji & Crowder, 1991; Intons-Peterson, 1992; Loftus, 1991; Neisser, 1991). It is concluded that researchers generally agree that both naturalistic and controlled methods should be used, but it is found (Poon, Welke, & Dudley, this volume) that researchers, in practice, use primarily controlled methods. The persistence of the debate may be due to unarticulated differences in the world views of the debaters.

Second, a parallel dialogue concerning the theoretical differences between laboratory and real-world tasks (e.g., Puckett, Reese, Cohen, & Pollina, 1991; Sinnott, 1989b) is discussed. Wide differences of opinion are found regarding the salient dimensions that distinguish everyday and laboratory tasks, although there has been little investigation of these dimensions.

Third, numerous mechanisms proposed to underlie everyday cognition are reviewed, including the creation and utilization of domains of expertise. Several other mechanisms are also identified that potentially help integrate findings across studies in this volume and across age periods of the life span.

Fourth, relationships between the mechanisms proposed to underlie practical and laboratory cognition are explored. The consensus seems to be that mechanisms are shared between areas. Finally, Puckett et al. (this volume) discuss several ways in which life-span cognitive research in both laboratory and everyday domains can compliment and enrich research in the other domain.

TAXONOMY AND METHODOLOGY IN EVERYDAY COGNITION

Poon et al.: Defining Everyday Cognition Methodologically

In the study of cognition, a controversy between proponents of controlled observation (e.g., Banaji & Crowder, 1989, 1991; Roediger, 1991) and those of naturalistic observation (e.g., Ceci & Bronfenbrenner, 1991; Neisser, 1978, 1991; see Puckett et al., this volume, for a review) has now existed for over a decade. In order to address this controversy, Poon, Welke, and Dudley (this volume) investigate the question, "What is everyday cognition?" in terms of methodology. They do so by observing the characteristics of published studies, sampling in a controlled manner from both everyday and laboratory domains, and assessing the characteristics of articles from each area. Controlled variables include the year of publication, number of articles from each source, and type of articles (only empirically based studies). Everyday studies are represented by studies in the volume edited by Gruneberg, Morris, and Sykes (1988) and by articles from the journal *Applied Cognitive Psychology.* Laboratory studies are represented by articles from the *Journal of Experimental Psychology: Learning, Memory, and Cognition.*

Poon et al. analyze the studies in terms of such methodological factors as the source of the research question, type of design used, type of stimuli used, type of conclusion, and whether theory is invoked in the design or interpretation of the study. They conclude on the basis of their meta-analysis that methodological differences between the two research areas are minimal, asking, "Where's the beef?" Then Poon et al. explain four sources of confusion that have contributed to the polarization of viewpoints between everyday and laboratory researchers. One source of confusion relates to their finding that, contrary to common belief, both camps of researchers utilize mainly laboratory research methods. Second, confusion exists in usage of the terms *representativeness of design, ecological validity,* and *functional validity.* Third, the proper priority of experimental control vis à vis other research considerations is widely misunderstood. Fourth, it appears that most traditional cognitive researchers share with most everyday cognitive researchers an "analytic" research model for which understanding is the goal; neither group as a whole appears to assume an "analogue" research model for which prediction is the goal (as is the case for medical and marketing researchers).

Poon et al. conclude their chapter by taking the position that the important question is not whether studies should be performed in the laboratory but rather when and why a particular setting should be selected. This conclusion reinforces the emerging consensus that both naturalistic and laboratory methods should be used to study cognition (e.g., Ceci & Bronfenbrenner, 1991; Intons-Peterson, 1992; Roediger, 1991; Winograd, this volume). Poon et al. then elaborate a useful list of factors that should be considered in making decisions about the type of research setting best suited for a particular research question.

Willis and Schaie: Defining Everyday Cognition Theoretically

Willis and Schaie (this volume) advocate and use sophisticated psychometric and experimental methods in the investigation of everyday cognition, essentially the same methods traditionally used in laboratory cognition. Although they mention naturalistic observation, the bulk of their research seems to utilize generally less naturalistic procedures such as interviews, thinking-aloud protocol analysis, paper-and-pencil tests, task simulations, and computer-interactive problems. Willis and Schaie address a host of methodological issues in their chapter, including sampling, validity, task analysis, and judging the adequacy of problem solution.

For Willis and Schaie, there is apparently no distinction between methodological strategies for everyday and laboratory cognition. In contrast, they do distinguish practical from laboratory cognition in such theoretical terms as the complexity and multidimensionality of problems (see also Ceci & Hembrooke, this volume). However, they assert that there is far less consensus regarding a number of other dimensions such as ill- vs. well-structured, despite the fact that some of the other contributors list such factors (e.g., Chapman, this volume; Sternberg, Wagner, & Okagaki, this volume). Willis and Schaie further observe that the theoretical dimensions that separate practical and laboratory cognition can be categorized on the basis of: (a) personal attributes, for example, Walters and Gardner's (1986) types of intelligence, with verbal and mathematical intelligence dominating academic cognition; (b) task properties, for example, Sinnott's (1989a) dual criterion for practical cognition, defined in terms of task frequency and task significance; and (c) task context, for example, Scribner's (1986) position that real-world tasks are embedded within the higher level goals of real life.

Willis and Schaie outline their longstanding and productive research program with old adults involving seven areas of everyday functioning that include telephone usage, shopping, and financial management. With respect to the seven areas, they have investigated what they see as the theoretical components or mechanisms of practical intelligence. Some of those elements listed are relevant skills, domain-specific knowledge, understanding of the present context, prior attitudes, and integration of the other components. What is the relationship of these mechanisms to those used in laboratory cognition? The answer may lie in Willis and Schaie's suggestion that processes are organized hierarchically, with unique combinations of basic processes organized to perform each type of task according to such factors as the demands of the situation and the individual's motivational state.

Winograd: Integrating Research Methods

Winograd (this volume) argues that researchers in everyday memory should increase the use of naturalistic methods while continuing to use laboratory methods. Although there have been numerous reasons offered in defense of naturalistic methods, Winograd emphasizes the rationale that many phenomena are difficult, if not impossible, to study with laboratory methods (see also Bahrick, 1991).

To make his point, Winograd marshalls numerous examples of research on real-world phenomena that have profited from the use of naturalistic methods. One of these is involuntary memory, which can apparently be loosely defined as memory retrieval in which no obvious external cues or internal volition have elicited the retrieval. By definition, then, one could not plan an involuntary retrieval event in a laboratory (or any other) setting. Another example is that of very-long-term memory. Even Banaji and Crowder (1991), two of the major proponents of laboratory methods, acknowledge that studying retention over a 50-year interval in the laboratory would be "vanishingly unrealistic." Among other phenomena that Winograd discusses is autobiographical memory, which might also be unrealistically difficult to study to any substantial extent in the laboratory.

Even while pointing out the necessity of using naturalistic methods in studying real-world memory problems, Winograd emphasizes the continuity between the laboratory and the real world. In order to document this continuity, he relates examples where principles derived in the laboratory have been useful in explaining everyday phenomena. In doing so, Winograd delineates several principles that suggest cognitive mechanisms. For example, he cites overloading and distribution of practice, principles uncovered in the laboratory, as explanations of recall of high school classmates. Perhaps it is the case that only in the field can enough data regarding certain phenomena be gathered to sufficiently describe them, but perhaps only in the laboratory can the underlying cognitive mechanisms be explicated. This position is not unlike that of Roediger (1991), Rubin (1989), and others. However, Winograd's chapter is possibly the most convincing documentation of research topics to date that demonstrates the need for diversity of methods.

THEORY AND METATHEORY FOR LIFE-SPAN EVERYDAY COGNITION

Sinnott: Theory and Metatheory in Everyday Cognition and Other Sciences

Sinnott (this volume) is concerned with conceptual and metaconceptual questions. First, she examines the metatheoretical framework used by many if not most everyday cognitive researchers and explores the kinds of questions whose investigation is facilitated by that framework. She suggests that the everyday framework benefits the investigation of such issues as cognitive processes rather than states, levels of analysis of cognitive phenomena, operation of basic cognitive mechanisms in the real world, the adaptability of cognitive processes, and cognitive growth in midlife and old age.

Sinnott then investigates the constructs that everyday cognitive research potentially shares with and can gain from an examination of other new areas of scientific inquiry. These areas and some of the shared principles include the contextuality of truth (shared with the new physics), the interdependence of parts, from which parts a functioning system emerges (from the new biology), the codetermination of multiple elements in a complex system of organized disorder (from chaos theory), and the hierarchical control of simpler thought processes by more complex modes (from postformal cognitive psychology).

Finally, she explores how the study of practical cognition relates to descriptions of life-span development in middle and old age. Using General Systems Theory (e.g., Laszlo, 1972) and principles borrowed from other sciences, Sinnott illustrates how practical cognition operates to shape the dialectics of adult development. For example, a young adult realizing that she must choose a career path from among many possible paths must nevertheless make a passionate commitment to the one choice or truth. In the process, she relinquishes the long-held illusion that there was only one truth possible and simultaneously discovers that truths are at least partially self-created.

Sinnott then briefly documents her extensive program of research involving everyday thinking in old adults, describing the multiple methods utilized and the multiple mechanisms investigated in her studies. She concludes on the basis of these studies that the mechanisms of everyday cognition develop reliably, that they can be manipulated and modeled, and that they are different from, but related to, those of traditional information processing.

Chapman: A Dialectical Theory of Everyday Cognition

The late Michael Chapman (this volume) outlined his dialectical theory of everyday reasoning, proposing that it includes not only internal problem solving and internal belief revision but also the everyday process of overt argumentation.

One implication of this position is that in everyday reasoning one concerns oneself with the speaker's intentions, in contrast to formal reasoning in which one does not. In everyday argumentation, people use maxims (Grice, 1975) or rules of conversation such as giving complete information (part of the maxim of quantity). In the absence of evidence (or formal training) to the contrary, it may automatically be assumed that a source of communication observes Grice's maxims. For example, the statement "If it rains, then the grass gets wet," followed by "The grass is wet" often leads people to the conclusion that "It must have rained." By the rules of formal reasoning, this conclusion is incorrect. But in the real world, a reasonable assumption according to the maxim of quantity is that if there were other reasons for the grass getting wet, then the source would have mentioned them. Because none were mentioned, the wetness must have been due to rain. This conclusion is perfectly reasonable in the real-world context according to Chapman.

The internalization of such overt everyday communication habits was said by Chapman to engender intrapsychic everyday reasoning processes. This process of internalization was called sociogenesis (e.g., Vygotsky, 1978). Theories of intrapsychic everyday reasoning have much in common with those of overt everyday discourse, supporting Chapman's conceptualization of the sociogenesis of everyday reasoning. Chapman reviewed theories of intrapsychic nonmonotonic reasoning (e.g., Ginsburg, 1987) in which assumptions are made in the absence of information to the contrary; according to Chapman these types of assumptions match those made in everyday argument. Other parallels exist, such as that between the case of an external argument between two people and the case of internal belief revision in which one is seen as arguing dialectically with oneself between two opposing views.

Formal reasoning was said by Chapman to differ from practical reasoning in its sociogenetic origins on the basis of education and other cultural factors. For example, formal methods learned in school may become internalized as formal thinking. Another requirement of formal reasoning, however, is the processing capacity necessary to generate and consider exhaustively all of the possibilities inherent in a problem. This requirement is absent in everyday reasoning, where reasonable assumptions suffice, and is an adaptive strategy for individuals who lack the necessary processing resources for formal reasoning. The development of formal and everyday reasoning and the attentional constraints of each type of reasoning suggest that the same mechanisms are used in the two domains, but that their organization is tailored to suit each context.

Concerning the observable differences between academic and everyday reasoning tasks, Chapman appears to accept Galotti's (1989) list (see also Sternberg et al., this volume) that includes such variables as whether established methods of solution are available (i.e., ill- vs. well-structured), whether all of the premises are provided, and whether one or more solutions exist (i.e., ill- vs. well-defined). This list recognizes some differences between task domains that are questioned by Willis and Schaie (this volume).

Finally, Chapman commented on the debate between scientists concerning the logic and rationality of scientific methods, and perhaps implicitly, the debate between researchers in laboratory and practical cognition. Chapman acknowledged that everyday reasoning is illogical, as defined by the rules of formal logic, but asserted that everyday reasoning is just as *rational* as formal reasoning. Furthermore, he stated that science always retains some measure of everyday reasoning as well as formal logic. Perhaps one implication of this observation is that researchers in laboratory and practical cognition should learn to integrate the reasoning modes and methods of the other camp with those of their own camp.

EMPIRICAL STUDIES OF EVERYDAY COGNITION ACROSS THE LIFE SPAN

Ceci and Hembrooke: Mechanisms of Everyday Infant Cognition

Ceci and Hembrooke (this volume) are concerned with everyday infant memory. First, they outline the reasons for the rising interest in this area. In the 1980s there was a rapid rise in calls for expert witness testimony in response to increasing numbers of children testifying in court cases on sexual abuse. Researchers in childhood memory came to realize, based on their formerly laboratory-bound research, that they could not answer many of the courtroom questions posed about everyday infant memory. Rather than abandoning the laboratory, however, Ceci and Hembrooke do an admirable job of integrating laboratory and everyday research findings in searching for cognitive mechanisms. Such integration of research methods is what Ceci himself has elsewhere advocated (Ceci & Bronfenbrenner, 1991).

Several themes run throughout Ceci and Hembrooke's chapter, the central one perhaps being the importance of context. The authors appear to argue here that richness of context is the most salient feature distinguishing between real-world and laboratory tasks (see also Willis & Schaie, this volume), at least for infants (Ceci & Liker, 1986, also argued for the importance of expertise in practical cognition in adults). For example, real-world tasks involve a greater variety of social and emotional contextual cues than do laboratory tasks. Ceci and Hembrooke suggest that context is possibly even more important in infants than in older children. To explain this putative finding, one interesting mechanism offered by Ceci and Hembrooke is that infants actually encode more about situations than adults, but infants' encoding may be less selective and hence likely to be based on many irrelevant cues so that retrieval might be more disrupted by changes in the external context.

Another theme addressed by Ceci and Hembrooke concerns the temporal properties of early memories. One general trend is that the earlier the age at which the memory is encoded and stored, the more difficult it is to demonstrate that the memory is retrievable over a given retention interval. But such phenomena as infantile amnesia in rats can be ameliorated by the presentation of reminders of the original training, and under certain conditions long-term retention of very early memories in children can be demonstrated with the careful reintroduction of the proper cues.

Ceci and Hembrooke also consider the level of consciousness likely to be associated with early memories. The earliest memories may not be available for conscious retrieval, possibly because they are data-driven, implicit memories (e.g., Roediger & Blaxton, 1987). Another possibility is that conscious retrieval of memories involving the self may be impossible before one year of age in children because self-representations may not have developed before that age (Howe & Courage, in press).

An example related by Ceci and Hembrooke illustrates the complex interactions of factors determining infant memory. The earliest age at which an event can be encoded and at which that encoding can later be retrieved as a conscious memory may depend not only on the reinstatement of the proper reminders but also on the emotional and social context of the original event. The earliest age of occurrence for the birth of a sibling resulting in a conscious memory might be 24 months; for a family moving to a new home it might be 36 months; and for the death of a relative it might be 48 months (Usher & Neisser, 1991). An integration of laboratory and naturalistic methods shows promise in unraveling these factors.

Walters et al.: Everyday Application of a Major Theory

Walters, Blythe, and White (this volume) report on the application of Gardner's (1983) theory of multiple intelligences to classroom instruction. The theory easily encompasses everyday cognition inasmuch as different combinations of intelligences are said to be used for different tasks. Traditional academic tasks are thought to rely heavily on linguistic and logical-mathematical intelligence, and real-world tasks are said to require all the types of intelligence, namely, linguistic, logical-mathematical, interpersonal, intrapersonal, spatial, bodily-kinesthetic, and musical.

Walters et al. describe how task context is important in eliciting the use of appropriate abilities. For example, students asked to predict the temperature of a mixture of 10° water plus more 10° water may say that the temperature will be 20°. But asked to predict the temperature of a mixture of cold water plus more cold water, the answer is likely to be cold water of the same temperature. In the first case, the task context misleads the students to use the wrong kind of skills for the problem. Walters et al. also describe the role of social context and intrapersonal understanding in applying intelligence intelligently. These aspects of the application of skills are incorporated by Walters et al. into their Practical Intelligence for Schools (PIFS) training modules, which are units designed to train various everyday schoolroom skills (see also Sternberg, Wagner, & Okagaki, this volume).

In the quasi-experimental study reported in their chapter, Walters et al. give pre- and posttests to students in experimental and control classes. Ten PIFS modules are infused into the experimental classes' regular curriculum. They report that over half of the PIFS modules led to reliable improvements in tested performance. Domain-specific units (e.g., word problems) tended to be successfully taught whereas domain-nonspecific units (e.g., taking notes) tended not to be. The authors attribute this pattern partly to the perceived greater relevance of the domain-specific units. Walters et al. go on to discuss the extension of this type of training to such everyday adult skills as family relations and financial management (see also Willis & Schaie, this volume).

Siegel et al.: Empirical Approaches to Everyday Adolescent Thinking

Siegel, Cuccaro, Parsons, Wall, and Weinberg (this volume) employ empirical approaches in exploring adolescents' everyday thinking about emotions and risk-taking. They first provide a theoretical and historical review of the study of adolescent emotions, which they treat as a part of the study of everyday social cognition. They make the point that adolescents are closer to their parents in terms of social cognitions than may first appear, and that arguing with parents may be a product primarily of adolescents' need to exercise their increased capacity to argue (see also Chapman, this volume, on the role of argument in cognitive development).

Siegel et al. describe an empirical study of emotions that involved a combination of methods. Initial exploratory interviews helped identify the dimensions of adolescent emotional experiences that they then included in a paper and pencil questionnaire. Adolescents were then asked to rate on Likert-type scales selected dimensions of emotions that they had experienced when in designated emotional states. These self-ratings were then subjected to factor analysis, revealing mood (positive vs. negative) and energy (high vs. low) factors. The authors state that the next step in their study will be multidimensional scaling. They summarize their methodological approach as starting with a description of the phenomenon via informal and pilot studies, developing a working model, assessing its heuristic utility, and moving toward a fine-grained analysis of a formal model. Pollina,

Greene, Tunick, and Puckett (1992, in press) have used a similar approach involving self-ratings of everyday memory and factor analysis in studies of adult cognition (see also Schooler & Loftus, this volume).

Next, in a study of risk-taking Siegel et al. interview adolescents and their parents on the perceived risks and benefits of smoking cigarettes and drinking alcohol. They find that adolescents are influenced by parents' attitudes and behaviors and conclude that "adolescents become increasingly aware of the benefits of smoking and drinking." Siegel et al. then propose everyday social cognitive mechanisms to explain such findings. They propose that adolescents perceive engaging in such hazardous activities as smoking, drinking, and unprotected sex as experience-seeking rather than as risk-taking. The lack of perceived risk in engaging in these kinds of activities may be related to adolescents' egocentric beliefs (e.g., Elkind, 1985) such as "it can't happen to me." The further elucidation of such mechanisms promises to be vitally important both practically and theoretically.

Schooler and Loftus: Delineating Mechanisms of Eyewitness Memory

At the conference on which this volume is based, Elizabeth Loftus commented on what is probably the most extensive and longstanding program of research in everyday cognition (e.g., Loftus, 1979). Before the movement in everyday cognition, her methods of studying eyewitness memory were criticized by laboratory researchers for not being sufficiently controlled. Since the everyday cognition movement began, her methods have been criticized by everyday researchers for being too controlled. In fact, Loftus and colleagues have tended to use fairly controlled methods, but so have the majority of researchers in both camps (Poon et al., this volume).

Schooler and Loftus (this volume) focus on the mechanisms of everyday eyewitness memory as typically investigated in a standard misinformation paradigm in which misleading questions are presented. One unifying concept in the chapter is the distinction between "immediate misinformation acceptance" and "delayed misinformation retrieval." This distinction is put to much use in explaining the results of demographic studies. Compared to adults, children appear to have poorer resistance to initial misinformation because of poorer recall of the original event. But children do not exhibit poorer accuracy upon later testing for the original information, suggesting to Schooler and Loftus that children also forget the misinformation that was initially accepted: one bad memory cancels another. Old adults have poorer accuracy than do young adults at the later testing. This suggests that old adults' retrieval of the original event is poorer than young adults', allowing them to accept and encode the misinformation, and on the final test, retrieval of the misinformation is selective in that its content but not its source is remembered. Thus the misinformation is fused with memory for the original event. These mechanisms are much like those posited in laboratory studies of cognitive aging (e.g., Kausler, 1991; Salthouse, 1991).

Schooler and Loftus then explore individual differences in cognition and personality as a "crucible of theory construction" (Underwood, 1975) in what can be described as construct validation of hunches derived from the demographic studies. For example, if memory capacity is responsible for developmental differences in eyewitness testimony, then individual differences in that capacity within a given age group should also be related to eyewitness memory (see Pollina et al., 1992, for a similar approach involving self-reported everyday memory ability and laboratory measures of memory performance). Schooler and Loftus' review reveals that immediate misinformation acceptance is associated with several factors such as self-esteem and memory performance. Delayed misinformation retrieval is associated with such factors as general intelligence, self-reported visual imagery ability, and self-reported general memory ability. Schooler and Loftus generate a number of hypotheses regarding the operation of laboratory-like mechanisms to explain individual differences in these real-world memory phenomena.

Sternberg et al.: Everyday Validation of a Major Theory

Sternberg, Wagner, and Okagaki (this volume) relate an impressive array of studies that validate the extension of Sternberg's (1985) triarchic theory of intelligence to everyday cognition in work and school. The theory encompasses three subtheories of intelligence: (1) componential (itself involving three classes of laboratory-type cognitive mechanisms: information processing, knowledge acquisition, and metacomponents); (2) experiential (experience and insight); and (3) contextual (involving three functions: adaptation to, selection of, and shaping of environments). Mechanisms of everyday/practical and laboratory/academic cognition apparently are seen to be parts of a single, larger cognitive system. The authors, however, recognize differences between everyday and academic tasks, providing a list similar to that of Galotti (1989; see also Chapman, this volume).

Sternberg's extension of the triarchic theory to the solution of real-world tasks is called tacit knowledge, which is defined as "practical know-how that usually is not directly taught or even openly expressed." Three aspects of tacit knowledge (managing oneself, managing others, and managing tasks) are proposed to exist for both local settings (i.e., strategies tailored to the immediate situation) and global settings (i.e., general strategies). These can in turn be considered with respect to both real and ideal jobs.

In order to validate the construct of tacit knowledge, Sternberg et al. use fairly controlled laboratory-like methods in settings of varying approximation to the real world. Tacit academic knowledge tests are validated for professors (Experiments 1 and 4 assessing discriminant validity) and students (Experiments 9 and

10 assessing convergent validity and analyzing an experiment with multivariate analysis of variance). Tacit knowledge for business managers is discriminant-validated (Experiments 2 and 5), cross-validated (Experiment 3), and concurrent-validated (Experiment 6). Tacit knowledge for salespeople is tested for discriminant validity (Experiment 7) and cross validity (Experiment 8). The authors conclude, among other things, that tacit knowledge does not correlate reliably with conventional intelligence and personality, that tacit knowledge among different areas nevertheless correlates moderately well (suggesting a "g" component for tacit knowledge), and that tacit knowledge correlates well with many meaningful variables including experience. Thus, even though tacit knowledge is viewed in the context of the triarchic theory as being part of a larger cognitive system utilizing the same componential mechanisms as traditional academic intelligence, the conclusion is that tacit knowledge is independent.

Finally, Sternberg et al. investigate (Experiment 11) the training of tacit knowledge in the Practical Intelligence for Schools (PIFS) program (see also Walters et al., this volume). Sternberg's group has responsibility for a core course consisting of relatively global tacit knowledge. In the PIFS program, Walters' group has responsibility for more specific (local) content courses, although Walters' units vary in specificity of content from what Sternberg would describe as relatively local to relatively global. Sternberg et al. report that their global tacit knowledge course is taught successfully in the PIFS program. In contrast, Walters et al. (this volume) report that their global (nonspecific) units are not taught reliably but that their local (specific) units are. The reasons for this discrepancy are not readily apparent. Whether or not global tacit knowledge can be trained, however, the studies reported by Sternberg et al. taken together strongly support the tacit knowledge construct.

CONCLUSIONS

The various theories, mechanisms, and findings regarding everyday cognition that are presented in this book indicate that it is a rich and vital area of research. Similar to the call made by Craik and Jennings (1992) with regard to cognitive aging and by Sinnott (this volume) with regard to everyday cognition, we believe that even broader integrations of the life-span everyday research efforts in this volume with those of other cognitive sciences and neurosciences will result in productive cross-fertilization and more rapid advancement in understanding human cognition.

REFERENCES

Bahrick, H. P. (1991). A speedy recovery from bankruptcy for ecological memory research. *American Psychologist, 46,* 76–77.

Banaji, M. R., & Crowder, R. G. (1989). The bankruptcy of everyday memory. *American Psychologist, 44,* 1185–1193.

Banaji, M. R., & Crowder, R. G. (1991). Some everyday thoughts on ecologically valid methods. *American Psychologist, 46,* 78–79.

Ceci, S. J., & Bronfenbrenner, U. (1991). On the demise of everyday memory. "The rumors of my death are much exaggerated" (Mark Twain). *American Psychologist, 46,* 27–31.

Ceci, S. J., & Liker, J. (1986). Academic and nonacademic intelligence: An experimental separation. In R. J. Sternberg & R. K. Wagner (Eds.), *Practical intelligence: Nature and origins of competence in the everyday world* (pp. 119–142). New York: Cambridge University Press.

Craik, F. I. M., & Jennings, J. M. (1992). Human memory. In F. I. M. Craik & T. A. Salthouse (Eds.), *The handbook of aging and cognition* (pp. 51–110). Hillsdale, NJ: Lawrence Erlbaum Associates.

Elkind, D. (1985). Egocentrism redux. *Developmental Review, 5,* 218–226.

Galotti, K. (1989). Approaches to studying formal and everyday reasoning. *Psychological Bulletin, 105,* 331–351.

Gardner, H. (1983). *Frames of mind: The theory of multiple intelligences.* New York: Basic.

Ginsburg, M. L. (1987). Introduction. In M. L. Ginsburg (Ed.), *Readings in nonmonotonic reasoning* (pp. 1–23). Los Altos, CA: Morgan Kaufmann.

Grice, H. P. (1975). Logic and conversation. In P. Cole & J. L. Morgan (Eds.), *Syntax and semantics: Vol. 3. Speech acts* (pp. 41–58). New York: Academic Press.

Gruneberg, M. M., Morris, P. E., & Sykes, R. N. (Eds.). (1988). *Practical aspects of memory: Current research and issues* (Vol. 1). New York: Wiley.

Howe, M. L., & Courage, M. L. (in press). On resolving the enigma of infantile amnesia. *Psychological Bulletin.*

Intons-Peterson, M. J. (1992). Brief introduction to the special issue on memory and cognition applied. *Memory and Cognition, 20,* 323–324.

Kausler, D. H. (1991). *Experimental psychology, cognition, and human aging* (2nd ed.). New York: Springer-Verlag.

Laszlo, E. (1972). *The relevance of general systems theory: Papers presented to Ludwig von Bertalanffy on his seventieth birthday.* New York: George Braziller.

Loftus, E. F. (1979). *Eyewitness testimony.* Cambridge, MA: Harvard University Press.

Loftus, E. F. (1991). The glitter of everyday memory . . . and the gold. *American Psychologist, 46,* 16–18.

Neisser, U. (1978). Memory: What are the important questions? In M. M. Gruneberg, P. E. Morris, & R. N. Sykes (Eds.), *Practical aspects of memory* (pp. 3–24). San Diego, CA: Academic Press.

Neisser, U. (1991). A case of misplaced nostalgia. *American Psychologist, 46,* 34–36.

Pollina, L. K., Greene, A. L., Tunick, R. H., & Puckett, J. M. (1992). Dimensions of everyday memory in young adulthood. *British Journal of Psychology, 83,* 305–321.

Pollina, L. K., Greene, A. L., Tunick, R. H., & Puckett, J. M. (in press). Dimensions of everyday memory in late adulthood. *Current Psychology.*

Puckett, J. M., Reese, H. W., Cohen, S. H., & Pollina, L. K. (1991). Age differences versus age deficits in laboratory tasks: The role of research in everyday cognition. In J. D. Sinnott & J. C. Cavanaugh (Eds.), *Bridging paradigms: Positive development in adulthood and cognitive aging* (pp. 113–130). New York: Praeger.

Roediger, H. L. (1991). They read an article? A commentary on the everyday memory controversy. *American Psychologist, 46,* 37–40.

Roediger, H. L., & Blaxton, T. A. (1987). Retrieval modes produce dissociations in memory for surface information. In D. S. Gorfein & R. R. Hoffman (Eds.), *Memory and cognitive processes: The Ebbinghaus Centennial Conference* (pp. 349–379). Hillsdale, NJ: Lawrence Erlbaum Associates.

Rubin, D. C. (1989). Introduction to Part I: The how, when, and why of studying everyday cognition. In L. W. Poon, D. C. Rubin, & B. A. Wilson (Eds.), *Everyday cognition in adulthood and late life* (pp. 3–27). Cambridge, England: Cambridge University Press.

Salthouse, T. A. (1991). *Theoretical perspectives on cognitive aging*. Hillsdale, NJ: Lawrence Erlbaum Associates.

Scribner, S. (1986). Thinking in action: Some characteristics of practical thought. In R. J. Sternberg & R. K. Wagner (Eds.), *Practical intelligence: Nature and origins of competence in the everyday world.* (pp. 13–30). New York: Cambridge University Press.

Sinnott, J. D. (1989a). An overview—if not a taxonomy—of "everyday problems" used in research. In J. D. Sinnott (Ed.), *Everyday problem solving: Theory and applications* (pp. 40–54). New York: Praeger.

Sinnott, J. D. (1989b). Background: About this book and the field of everyday problem solving. In J. D. Sinnott (Ed.), *Everyday problem solving: Theory and applications* (pp. 1–6). New York: Praeger.

Sternberg, R. J. (1985). *Beyond IQ: A triarchic theory of human intelligence*. New York: Cambridge University Press.

Underwood, B. J. (1975). Individual differences as a crucible in theory construction. *American Psychologist, 30,* 128–134.

Usher, J., & Neisser, U. (in press). Childhood amnesia in the recall of four target events. *Journal of Experimental Psychology: General.*

Vygotsky, L. (1978). *Mind in society.* Cambridge, MA: Harvard University Press.

Walters, J. M., & Gardner, H. E. (1986). The theory of multiple intelligences: Some issues and answers. In R. J. Sternberg & R. K. Wagner (Eds.), *Practical intelligence: Nature and origins of competence in the everyday world* (pp. 163–182). New York: Cambridge University Press.

Author Index

Subject Index